A MANUAL OF INVERTEBRATE PALEONTOLOGY

by
JUDITH C. DYER
FREDERICK R. SCHRAM
San Diego Natural History Museum

ILLUSTRATED BY
PATRICIA C. AMTOWER

ISBN 978-0-87563-237-7

Published by
Stipes
PUBLISHING
www.stipes.com

CONTENTS

PREFACE

This manual was written to satisfy a frustrating condition we have encountered in our academic experience, the lack of a beginner's guide to the identification of fossils. Our intention has been to provide an introductory study guide or outline of paleontology. We have not produced an extensive text, since there are many books, among them the *Treatise on Invertebrate Paleontology*, which more adequately cover the broad subject area.

This manual is instead intended to supplement such major works by providing an easy to use guide to the identification and the characteristic features of the major invertebrate fossils. We have tried to address the problems which a beginning student encounters in the field or a paleontology laboratory, namely, "What am I looking at?", and "What am I supposed to be looking for?" Many are capable of finding their own way, but in our experience they find it more quickly and effectively if offered some guidance.

The wealth of fossil material that potentially could have been included in this manual presented difficulties, both of selection and scope. The manual is not encyclopedic nor was it intended to be. It emphasizes what we feel are a basic array of fossil types to which beginning students should be exposed. Specimens of these taxa are available from supply houses such as Wards, Rochester, New York, or Carolina Biological Supply, Burlington, North Carolina. However, this is not to discourage instructors from including materials from their local paleofauna to supplement the fossils presented here. Among our objectives for this manual is the hope that it will help standardize the material presented to beginning students of basic paleontology and, thus, achieve some equivalence with the systematic approach of other scientific disciplines. We naturally encourage users of this manual to offer constructive suggestions for the improvement of future editions.

INTRODUCTION

This manual contains chapters on various groups of fossils which are important for a basic understanding of past life on Earth. Although, many approaches are possible, we have chosen a taxonomic and morphological view. We feel that this offers the beginning student the most obvious framework within which to organize his growing knowledge.

Each chapter contains a short discussion of the basic morphological features of the relevant group. Important words or characteristics are underlined and the reader is encouraged to find and identify each of these on the specimens made available for study. As a learning aid, the reader should label the drawings included in each chapter with the representative characteristics of each fossil group.

A subsection of each chapter deals with basic variants within each group of fossils. Specimens, or representative local materials, should be examined in detail and related to the basic morphology outlined in the first part of each chapter. Learning and understanding will be greatly enhanced if each item is analyzed in comparison with the other materials in that group. The objective should be to acquire a practical knowlege of basic form and comparative variation. If the manual is being used as a laboratory text, this knowledge may then be tested in practical examinations using fossil specimens.

A basic taxonomy is provided in Chapter 2 to relate each group to the total array of organic diversity.

Chapter 1: FOSSILIZATION

A. Discussion

Fossilization is the process that preserves evidence of past life on earth. Usually only the hard inorganic parts of an animal, such as its bones, teeth, and shell are preserved. The soft tissues typically decay too rapidly to leave evidence of their existence. However, under unique circumstances the soft parts of animals, as well as plants, may be preserved, giving us an unusual record of previous life on the earth.

Hard Parts

Bones, teeth, and shell are fossilized by a number of processes, several of which may be considered together. These are permineralization (petrifaction), recrystallization, and replacement.

Permineralization or petrification occurs when minerals in solution are deposited in the intercellular spaces or pores of an animal or plant. Two types of material most often found preserved by this method are bone and wood. Mesozoic dinosaur bones from Wyoming and Colorado as well as petrified wood from Arizona are prominent examples of such permineralized fossils.

Permineralization occurred when the bone or wood became saturated with mineral rich water. In time, the minerals precipitated out of solution and gradually filled the small intercellular cavities. The original structure surrounding the cavities was retained intact or altered by replacement or recrystallization. Under ideal conditions, the minerals preserve many of the details of the original structure, even though the composition of the fossil is quite different from that of the once living organism. In handling such fossils the student should note that permineralized specimens are much heavier than the original material would have been.

Recrystallization results from the natural tendency of all minerals to seek their most stable form. The shells of many invertebrates are composed of aragonite which over time will recrystallize as calcite. Both minerals are composed of the same chemical compound, calcium carbonate, but calcite is the more stable form of the two. Many shells show this type of fossilization.

Replacement is a type of fossilization which takes place when the original material of an organism is dissolved and replaced by other mineral compounds, most commonly silica, pyrite, or calcium carbonate. Pyritized brachiopods, frequently found in the Paleozoic rocks of the midwestern United States, or silicified brachiopods from the Permian of west Texas are notable examples of the process of replacement.

Soft Parts

The soft organic parts of an organism are fossilized only when a special situation prevents the normal course of decay. Freezing hinders decay by slowing bacterial reproduction and activity. Low temperatures, which prevented bacterial action, allowed the preservation of certain Pleistocene fauna, including frozen specimens of the extinct Siberian mammoth. When thawed, the mammoth's flesh was found to be quite fresh even though over 10,000 years old.

Submersion in tar, oil, or resin prevents decay of both soft and hard parts. Microorganisms cannot survive in such an environment and thus do not destroy animal or plant tissue. Ancient Baltic amber frequently contains many species of fossil insects which during their life became entrapped in the sap of conifers and preserved for millions of years in perfect condition. The La Brea tar pits of Los Angeles, California, have yielded preserved skeletal specimens of numerous Pleistocene vertebrates which became mired in the oil as they sought to drink the water floating on the surface of tarry pools.

Sometimes the soft parts are preserved by a special form of fossilization called carbonization. The carbonized remains of plants and animals result from the alteration of the organic material by decay, pressure, and heat leaving only a thin film of carbon where once had been a whole organism. Graptolites, which puzzled paleontologists for years because of their resemblance to pencil marks on slate, are almost exclusively known by their carbon film remains. The carbonization process is also responsible for the well known fossil fish of the Green River area of Wyoming. In these fossils an extraordinary amount of detail has been preserved.

Occasionally the fossil record of an organism is not found in the preserved remains of its hard or soft parts but in indirect evidence such as casts or molds. A snail may have been buried in the mud at the bottom of a sea or lake and in time the mud hardened. Eventually the snail's body decayed and the shell slowly dissolved leaving only a cavity in the hardened sediments. The cavity that remains is known as a mold and will often preserve the exact external details of the shell. If the mold had filled with sediments or minerals which also hardened, an exact replica or cast of the shell would be preserved. Not only are organisms preserved as casts and molds but also preserved are signs of their activities such as trails, footprints, toothmarks, and borings. These are known as trace fossils.

Chapter 1: Fossilization

Generally fossils, especially invertebrate fossils, are found in sedimentary rocks which have been deposited by water. The best sedimentary material for preservation is fine grained sand, silt, or clay. These particles are small enough so that their weight and abrasive properties are relatively slight. They gently entrap animal or plant remains as they slowly settle to the bottom of pools, lakes, seas, and oceans. As the sediment accumulates and compacts, it provides protection to the organisms lying in the accumulation and enables fossilization to occur.

Types of rocks other than sedimentary rocks, such as metamorphic or igneous rocks, usually do not contain fossils. The tremendous pressure and heat required in their formation generally destroys any potential fossils in their midst.

Samples of each fossilization process should be examined in order to gain first-hand experience in recognizing and in understanding them.

NOTES

Chapter 2:
BRIEF OUTLINE of the CLASSIFICATION of ANIMALS

Taxonomy is a dynamic discipline. Classifications are not musty old relics of some ordered but stodgy mind. Rather, classifications are state-of-the-art devices for organizing our knowledge concerning living things. To the dismay of students, and frequently instructors, it is not uncommon to find reputable textbooks not agreeing in the taxonomic schemes they use. These in turn may not agree with what active experts in their fields of study may use.

This chapter utilizes the latest schemes for particular groups. We have tried to give a complete classification down to the subclass level where applicable. In a few instances we have included some of the hundreds of orders where it is particularly instructive. Our object is to demonstrate to students the range of life in the past in comparison with the Recent or modern biota and to illustrate how relatively little material is covered at the introductory level compared to the total possible range of fossils available in the world.

Difficulties are encountered in trying to sort out what is and is not an animal. Biologists today recognize five kingdoms of organisms:

Kingdom Monera (bacteria, blue-green algae)
Kingdom Protista (protozoans, protophytans)
Kingdom Fungi (mushrooms, molds, rusts)
Kingdom Plantae (algae, plants)
Kingdom Animalia (animals)

General courses in paleontology typically cover forms assignable to three of these five kingdoms: protistans, plants, and animals. This manual is concerned for the most part with protozoans and invertebrate animals and these are emphasized in the following taxonomy.

Finally, several sections of this classification are admittedly controversial. We have deliberately tried to incorporate the latest taxonomies available in respective phyla. Since people have mistaken views of what a classification is (some kind of sacred cow or codified rubric), they are always loath to give up "old familiar friends". Three groups are especially controversial.
1) There is much research currently undertaken on Cambrian molluscan types. New species are being discovered at an astounding rate, and resultant disagreements among prominent authorities such as Pojeta, Runegar, and Yochelson are not

uncommon. The molluscan classification presented here is being outdated by new discoveries even as the student reads these words. 2) Much analysis of arthropod functional morphology and embryology has occured in the 1960s and 1970s. This has resulted in our understanding how diverse a group arthropods are. Authorities like Anderson, Manton, Schram, and Whittington feel this diversity should be finally and openly admitted by placing the distinctive arthropodous types into four separate phlya. 3) Echinoderm taxonomy has always been in a state of flux. Succeeding generations have tried to grapple with the problem of arranging echinoderms into natural biological units. Until recently, the overwhelming number of extinct classes in this phylum forced people to focus strictly on external geometry and topology of echinoderm types. Concerted efforts to understand internal anatomy have resulted in a tentative classification by Bell and Haugh to reflect functional morphology and embryology.

These controversial taxonomies are already achieving their aims. They are stimulating new lines of thought and research to test whether the relationship these classifications suggest are appropriate and warrant continued use. In the meantime, students should not be dismayed by use of different taxonomies between books, or even succeeding editions of this laboratory manual. Disagreements merely reflect progress in a dynamic field of study.

Chapter 2: Classification

Kingdom Protista [Precambrian-Recent]

(non animal-like phyla: Euglenophyta, Chrysophyta, Pyrrophyta, Hypochytridiomycota, Plasmidiophoromycota)

(animal-like phyla)

Phylum Sporozoa [Recent]

Phylum Cnidosporidia [Recent]

Phylum Sarcomastigophora [Cambrian-Recent]

Class Mastigophora [Recent]

Class Opalinata [Recent]

Class Rhizopoda [Cambrian-Recent]

Class Actinopoda [Ordovician-Recent]

Phylum Ciliophora [Ordovician-Recent]

Class Holotrichia [Recent]

Class Peritricha [Recent]

Class Suctoria [Recent]

Class Spirotrichia [Ordovician-Recent]

Kingdom Animalia [Precambrian-Recent]

Subkingdom Agnotozoa

Phylum Mesozoa [Recent]

Subkingdom Parazoa

Phylum Porifera [Cambrian-Recent]

Class Hexactinellida [Cambrian-Recent]

Class Heteractinida [Cambrian-Permian]

Class Calcarea [Cambrian-Recent]

Class Demospongia [Cambrian-Recent]

Chapter 2: Classification

Phylum Archaeocyatha [Cambrian]

Class Regulares

Class Irregulares

Subkingdom Metazoa

Phylum Coelenterata [Late Precambrian-Recent]

Class Protomedusae [Late Precambrian-Ordovician]

Class Hydrozoa [Precambrian-Recent]

Order Stromatoporoidea [Cambrian-Cretaceous]

Class Scyphozoa [Late Precambrian-Recent]

Subclass Scyphomedusa [Late Precambrian-Recent]

Subclass Conulata [Middle Cambrian-Triassic]

Class Anthozoa [Ordovician-Recent]

Subclass Ceriantipatheria [Miocene-Recent]

Subclass Octocorallia [Permian-Recent]

Subclass Zoantharia [Ordovician-Recent]

Infraclass Tabulata [Ordovician-Permian]

Infraclass Heliolitoida [Ordovician-Devonian]

Infraclass Rugosa [Ordovician-Permian]

Infraclass Heterocorallia [Late Carboniferous]

Infraclass Scleractinia [Triassic-Recent]

Infraclass Corallimorpharia [Recent]

Infraclass Zoanthinaria [Recent]

Infraclass Actiniaria [Recent]

Phylum Ctenophora [Recent]

Chapter 2: Classification

Phylum Platyhelminthes [Recent]

- Class Turbellaria
- Class Monogenea
- Class Trematoda
- Class Cestoda

Phylum Nemertea [(Middle Cambrian ?) Carboniferous-Recent]

Phylum Acanthocephala [Recent]

Phylum Rotifera [Recent]

Phylum Entoprocta [Recent]

Phylum Gastrotricha [Recent]

Phylum Nematomorpha [Recent]

Phylum Kinorhyncha [Recent]

Phylum Nematoda [Carboniferous-Recent]

Phylum Priapulida [Cambrian-Recent]

Phylum Sipuncula [Recent]

Phylum Echiura [Carboniferous-Recent]

Phylum Annelida [Late Precambrian-Recent]

- Class Polychaeta [Late Precambrian-Recent]
- Class Myzostomaria [Recent]
- Class Oligochaeta [Recent]
- Class Hirudinoidea [Recent]

Phylum Uniramia [(Middle Cambrian ?) Devonian-Recent]

- Subphylum Onychophora [(Middle Cambrian ?) Carboniferous-Recent]
- Subphylum Euthycarcinoidea [Carboniferous-Triassic]

Chapter 2: Classification

Subphylum Myriapoda [Silurian-Recent]

- Class Archipolypoda [Silurian-Carboniferous]
- Class Diplopoda [Carboniferous-Recent]
- Class Chilopoda [Cretaceous-Recent]
- Class Symphyla [Oligocene-Recent]
- Class Pauropoda [Recent]
- Class Arthropleurida [Carboniferous]

Subphylum Hexopoda [Devonian-Recent]

- Class Protura [Recent]
- Class Collembola [Devonian-Recent]
- Class Diplura [Oligocene-Recent]
- Class Thysanura [Carboniferous-Recent]
- Class Pterygota [Carboniferous-Recent]

Phylum Crustacea [Cambrian-Recent]

- Class Remipedia [Recent]
- Class Cephalocarida [Recent]
- Class Branchiopoda [Devonian-Recent]
- Class Ostracoda [Cambrian-Recent]
- Class Maxillopoda [Cambrian-Recent]
  - Subclass Copepoda [Cretaceous-Recent]
  - Subclass Mystacocarida [Recent]
  - Subclass Branchiura [Recent]
  - Subclass Cirripedia [Cambrian-Recent]

Chapter 2: Classification

Class Malacostraca [Cambrian-Recent]

Subclass Phyllocarida [Cambrian-Recent]

Subclass Hoplocarida [Devonian-Recent]

Subclass Eumalacostraca [Devonian-Recent]

Phylum Cheliceriformes [Cambrian-Recent]

Subphylum Chelicerata [Cambrian-Recent]

Class Merostomata [Cambrian-Recent]

Subclass Xiphosurida [Cambrian-Recent]

Subclass Eurypterida [Ordovician-Permian]

Class Arachnida [Silurian-Recent]

Subphylum Pycnogonida [Devonian-Recent]

Phylum Trilobitomorpha [Cambrian-Triassic]

Class Trilobitoidea [Cambrian-Triassic]

Class Trilobita [Cambrian-Permian]

Order Agnostida [Cambrian-Ordovician]

Order Redlichiida [Cambrian]

Order Corynexochida [Cambrian]

Order Ptychopariida [Cambrian-Permian]

Order Phacopina [Cambrian-Devonian]

Order Lichida [Ordovician-Devonian]

Order Odontopleurida [Cambrian-Devonian]

Phylum Mollusca [Cambrian-Recent]

Class Stenothecoida [Cambrian]

Class Hyolithida [Cambrian-Permian]

Class Rostrochonchia [Cambrian-Permian]

Class Pelecypoda [(Cambrian ?) Ordovician-Recent]

Subclass Palaeotaxodonta [Ordovician-Recent]

Subclass Isofilibranchia [Ordovician-Recent]

Sublcass Pteriomorpha [Ordovician-Recent]

Subclass Heteroconchia [Ordovician-Recent]

Subclass Anomalodesmata [Ordovician-Recent]

Class Scaphopoda [Devonian-Recent]

Class Aplacophora [Recent]

Class Polyplacophora [(Upper Cambrian ?) Ordovician-Recent]

Class Monoplacophora [Cambrian-Recent]

Class Cephalopoda [Cambrian-Recent]

Subclass Ectocochlia [Cambrian-Recent]

Superorder Nautiloidea [Cambrian-Recent]

Superorder Ammonoidea [Ordovician-Cretaceous]

Subclass Coleoidea [Carboniferous-Recent]

Class Gastropoda [Cambrian-Recent]

Subclass Prosobranchia [Cambrian-Recent]

Subclass Opisthobranchia [Devonian-Recent]

Subclass Pulmonata [Carboniferous-Recent]

Phylum Phoronida [Recent]

Phylum Bryozoa [Ordovician-Recent]

Class Stenolaemata [Ordovician-Recent]

Order Cryptostomata [Ordovician-Permian]

Order Cyclostomata [Ordovician-Recent]

Order Cystoporata [Devonian]

Order Trepostomata [Ordovician-Permian]

Class Gymnolaemata [Ordovician-Recent]

Order Ctenostomata [Ordovician-Recent]

Order Cheilostomata [Jurassic-Recent]

Class Phylactolaemata [Recent]

Phylum Brachiopoda [Cambrian-Recent]

Class Inarticulata [Cambrian-Recent]

Class Articulata [Cambrian-Recent]

Order Orthida [Cambrian-Permian]

Order Strophomenida [Ordovician-Jurassic]

Order Pentamerida [Cambrian-Permian]

Order Rhynchonellida [Ordovician-Recent]

Order Spiriferida [Ordovician-Jurassic]

Order Atrypida [Ordovician-Triassic]

Order Terebratulida [Devonian-Recent]

Phylum Echinodermata [Cambrian-Recent]

Subphylum I

Class Edrioblastoidea [Ordovician]

Class Crinoidea [Ordovician-Recent]

Class Edrioasteroidea [Cambrian-Carboniferous]

Class Helicoplacoidea [Cambrian]

Subphylum II

Class Blastoidea [Silurian-Permian]

Class Parablastoidea [Ordovician]

Class Diploporita [Ordovician-Devonian]

Class Rhombifera [Ordovician-Devonian]

Class Eocrinoidea [Cambrian-Silurian]

Class Paracrinoidea [Ordovician]

Class Ctenocystoidea [Cambrian]

Class Homostelea [Cambrian]

Class Homoiostelea [Cambrian-Devonian]

Class Stylophora [Cambrian-Devonian]

Subphylum III

Class Echinoidea [Ordovician-Recent]

Class Stelleroidea [Ordovician-Recent]

Class Holothuroidea [Devonian-Recent]

Subphylum Uncertain

Class Cyclocystoidea [Ordovician-Devonian]

Class Ophiocystoidea [Ordovician-Devonian]

Class Cycloidea [Cambrian]

Class Cymoidea [Cambrian]

Phylum Pogonophora [(Oligocene ?) Recent]

Phylum Hemichordata

Class Graptolithina [Cambrian-Mississippian]

Class Pterobranchia [Ordovician-Recent]

Class Enteropneusta [Recent]

Phylum Chordata [Cambrian-Recent]

Subphylum Urochordata [Recent]

Subphylum Cephalochordata [Carboniferous-Recent]

Subphylum Vertebrata [Cambrian-Recent]

Superclass Agnatha [Cambrian-Recent]

Class Monorhina [Recent]

Class Diplorhina [Cambrian-Recent]

Chapter 2: Classification

Superclass Gnathostomata [Silurian-Recent]

Class Placodermi [Devonian-Carboniferous]

Class Acanthodii [Silurian-Permian]

Class Chondrichthyes [Devonian-Recent]

Class Osteichthyes [Devonian-Recent]

Class Sarcopterygii [Devonian-Recent]

Class Amphibia [Devonian-Recent]

Class Reptilia [Carboniferous-Recent] *

Class Aves [Jurassic-Recent] *

Class Mammalia [Triassic-Recent] *

*It is widely agreed these are grades of organization and not realistic taxonomic categories. However, there is yet no consensus on a reasonable classification of monophyletic taxa to replace these classic classes.

For pedagogical purposes we also include below, a classification system that is similar to that used in other introductory paleontological and biological texts. This is not to negate the detailed taxonomy with geological ranges provided above. But the scheme below has some advantages in that it is familiar and perhaps somewhat easier to use. This simplified scheme is used in the group chapters.

Chapter 2: Classification

BRIEF OUTLINE OF THE CLASSIFICATION OF COMMON INVERTEBRATE FOSSILS

Kingdom Protista [Precambrian-Recent]

- Phylum Protozoa [Precambrian-Recent]
  - Class Sarcodina [Cambrian-Recent]
    - Order Foraminiferida [Cambrian-Recent]
      - Suborder Textulariina [Cambrian-Recent]
      - Suborder Fusulinina [Ordovician-Triassic]
      - Suborder Miliolina [Carboniferous-Recent]
      - Suborder Rotaliina [Permian-Recent]
    - Order Radiolaria [Ordovician-Recent]
- Phylum Chrysophyta [Cretaceous-Recent]

Kingdom Animalia [Precambrian-Recent]

- Phylum Porifera [Cambrian-Recent]
- Phylum Archaeocyatha [Cambrian]
- Phylum Coelenterata [Precambrian-Recent]
  - Class Hydrozoa [Precambrian-Recent]
    - Order Stromatoporoidea [Cambrian-Cretaceous]
  - Class Scyphozoa [Precambrian-Recent]
    - Subclass Conulata [Middle Cambrian-Triassic]
  - Class Anthozoa [Ordovician-Recent]
    - Order Rugosa [Ordovician-Permian]
    - Order Tabulata [Ordovician-Permian]
    - Order Scleractinia [Triassic-Recent]
- Phylum Bryozoa [Ordovician-Recent]
  - Order Ctenostomata [Ordovician-Recent]
  - Order Cryptostomata [Ordovician-Permian]

Order Cyclostomata [Ordovician-Recent]

Order Trepostomata [Ordovician-Permian]

Order Cheilostomata [Jurassic-Recent]

Phylum Brachiopoda [Cambrian-Recent]

Class Articulata [Cambrian-Recent]

Class Inarticulata [Cambrian-Recent]

Phylum Mollusca [Cambrian-Recent]

Class Monoplacophora [Cambrian-Recent]

Class Polyplacophora [Ordovician-Recent]

Class Scaphopoda [Devonian-Recent]

Class Gastropoda [Carboniferous-Recent]

Class Bivalvia [Ordovician-Recent]

Class Cephalopoda [Cambrian-Recent]

Order Nautiloidea [Cambrian-Recent]

Order Ammonoidea [Devonian-Cretaceous]

Order Coleoidea [Carboniferous-Recent]

Suborder Belemnoidea [Permian-Cretaceous]

Phylum Trilobitomorpha [Cambrian-Triassic]

Class Trilobita [Cambrian-Permian]

Phylum Cheliceriformes [Cambrian-Recent]

Class Merostomata [Cambrian-Recent]

Phylum Crustacea [Cambrian-Recent]

Class Ostracoda [Cambrian-Recent]

Class Cirripedia [Cambrian-Recent]

Chapter 2: Classification

Phylum Echinodermata [Cambrian-Recent]

Subphylum Homalozoa [Cambrian-Devonian]

Subphylum Crinozoa [Cambrian-Recent]

Class Cystoidea [Ordovician-Devonian]

Class Blastoidea [Silurian-Permian]

Class Crinoidea [Ordovician-Recent]

Subphylum Asterozoa [Ordovician-Recent]

Class Asteroidea [Ordovician-Recent]

Class Ophiuroidea [Ordovician-Recent]

Subphylum Echinozoa [Cambrian-Recent]

Class Helicoplacoidea [Cambrian]

Class Edrioasteroidea [Cambrian-Mississippian]

Class Holothuroidea [Devonian-Recent]

Class Echinoidea [Ordovician-Recent]

Phylum Hemichordata [Cambrian-Recent]

Class Graptolithina [Cambrian-Mississippian]

Order Dendroidea [Cambrian-Mississippian]

Order Graptoloidea [Ordovician-Devonian]

Phylum Conodonta [Cambrian-Triassic]

Kingdom Plantae

Phylum Chlorophyta

(Receptaculitids) [Ordovician-Devonian]

NOTES

Chapter 3:
PROTOZOA and DIATOMS

A. Discussion

PROTOZOA

The phylum Protozoa is comprised of single celled organisms which live throughout the world wherever water is present.

Each protozoan consists of cytoplasm surrounded by a cellular membrane. Within the cytoplasm are a nucleus, containing genetic information, and structures called organelles which perform different cellular functions. One of the organelles is the contractile vacuole, which functions as a pump regulating the amount of water in the cell. Other organelles are the flagella, cilia, and pseudopodia which are all responsible for food getting and locomotion. They also provide the basis for classifying the Protozoa into their taxonomic categories. Protozoa in Class Flagellata have flagella while those in Class Sarcodina have pseudopodia.

Chapter 3: Protozoa and Diatoms

Food is variously obtained by absorption, photosynthesis, or ingestion of particles and smaller organisms, with digestion taking place in an organelle called the food vacuole. Waste products are simply diffused through the cell membrane to the outside. Respiration is the result of the diffusion of dissolved oxygen across the cell membrane from the watery medium to the inside of the cell.

For Protozoa, reproduction has several possibilities. It can be asexual by simple cell division into either equal or unequal daughter cells or a related process called schizogony, whereby a multinucleate cell gives rise to many uninucleate daughter cells. Reproduction may also be sexual by the simple temporary union of two cells and the subsequent tranfer of genetic information, a process known as conjugation. Sexual reproduction may also occur as the result of the sexual fusion of two gametes. In addition to their various reproductive capabilities, Protozoa are also capable of forming cysts, a device which enables them to withstand unfavorable environmental conditions for substantial periods of time and endure fossilization.

Foraminiferida

Of importance to paleontologists are the Protozoa of Class Sarcodina, which have pseudopodia for both feeding and locomotion. They have few organelles and morphologically are among the simplest Protozoa. Many Sarcodina, such as members of the order Foraminiferida, have tests or shells of calcium carbonate in diverse shapes and designs which are characteristic for each species. Foraminiferida have a complex reproductive cycle incorporating both sexual and asexual phases. During the sexual phase the tests have a larger proloculus or initial chamber and are termed megalospheric. During the asexual phase, the proloculum is small and the test is designated microspheric.

There are several suborders of Foraminiferida which are of great interest to geologists as well as paleontologists. In the suborder Textulariina, the protozoan gathers bits of sand and shell from its environment and secretes calcium carbonate to cement them together to form a somewhat granular appearing shell. Its shape is variable including uniserial, biserial, and planispiral forms.

The suborders Miliolina, Fusulinina, and Rotaliina secrete the entire shell. Miliolina shells are imperforate (lacking pores) and have a porcelaneous appearance. Suborder Fusulinina, an extinct group, have shells which resemble grains of wheat. In cross section, the internal chambers have septa between them and are elongated parallel to the coiling axis. Tests of the suborder Rotaliina are perforate (have pores) and are characterized by a hyaline or glassy appearance. The shape varies, depending on the species and whether the protozoan is benthonic or planktonic. Benthonic Rotaliina tend to be very regularly shaped in a spiral or planispiral while planktonic Rotaliina have tests of a globular or spherical shape. The tests of all the suborders are variable and may include elaborate appendages.

Because the tests of many species of Foraminiferida have been abundantly preserved in the earth's strata over a wide geographical range, the forams are useful index fossils for the dating of sedimentary rocks. Geologists have made use of the information that foram tests reveal in their search for oil throughout the world. Foram tests have also been responsible in part for the accurate dating of ocean floor core samples collected by the Deep Sea Drilling Project.

In addition to dating information, Foraminiferida provide important information about paleoevironments. The form and composition of tests of certain species are modified by temperature and salinity. Other species are restricted to particular sediment types and water depths. Thus, the occurrence of any of these species in a particular sedimentary formation can indicate much about the conditions that existed at the time that species of Foraminiferida was alive.

Chapter 3: Protozoa and Diatoms

Radiolaria

Radiolaria are yet another order of the Class Sarcodina. They differ from Foraminiferida in that their tests are siliceous rather than calcareous, giving them a very glassy appearance. The shape is usually basket-like or spherical and commonly has spines which can be very elaborate. Radiolaria also differ from forams in that there are no fresh water species nor are there any benthonic species. They are all planktonic and inhabit many different depths in seas throughout the world. Their fossil history is very long, extending from Ordovician to recent times; however, their usefulness as biostratigraphic indicators is somewhat limited because their small size and great diversity have made them difficult to study.

DIATOMS

Diatoms are members of the phylum Chrysophyta. They are very like Protozoa in most respects but differ by having a rigid cell wall as well as a cell membrane. Diatoms can ingest food but are also photosynthetic. They are a very important primary source of nutrition in the food chain; and in the past history of the planet were crucial in producing the oxygen in our atmosphere. Diatoms resemble Radiolaria in having siliceous tests but the shape differs in that diatom tests resemple pill boxes in which the lid overlaps the bottom. They are distributed worldwide in all kinds of water and most soils. Geologically, diatoms are important as the component of diatomite or diatomaceous earth, which has many commercial uses in paints, plastics, polishing powders, and other products.

B. Synopsis

Phylum Protozoa [Precambrian-Recent]

- Single cells with cell membrane, cytoplasm, nucleus, organelles
- Motile by means of cilia, flagella, or pseudopodia
- Asexual and sexual reproduction; some genera form cysts
- Distributed worldwide wherever water is present

Class Sarcodina [Cambrian-Recent]

- Pseudopodia

Order Foraminiferida [Cambrian-Recent]

- Calcium carbonate test
- Microspheric and megalospheric forms of test
- Useful geological indicators

Suborder Textulariina [Cambrian-Recent]

- Test made of bits of shell and sand cemented together

Suborder Fusulinina [Ordovician-Triassic]

- Test calcareous of granular or coarse texture

Suborder Miliolina [Carboniferous-Recent]

- Test calcareous, porcelaneous, imperforate (no pores)

Suborder Rotaliina [Permian-Recent]

Test calcareous, hyaline, translucent, perforate (pores present)

Benthonic tests commonly spiral, small; planktonic tests globular, inflated

Order Radiolaria [Cambrian-Recent]

Siliceous test (made of silica), glassy, delicate, basket-like or spherical, spines common

Planktonic, marine

Phylum Chrysophyta [Cretaceous-Recent]

Common name, diatoms

Siliceous test, pillbox shape

Photosynthetic

Worldwide distribution in soils and water

C. Representative Groups

Textulariina [Cambrian-Recent]

Fusulinina [Ordovician-Triassic]

Miliolina [Carboniferous-Recent]

Rotaliina [Permian-Recent]

Radiolaria [Cambrian-Recent]

Chrysophyta [Cretaceous-Recent]

PROTOZOAN MORPHOLOGY

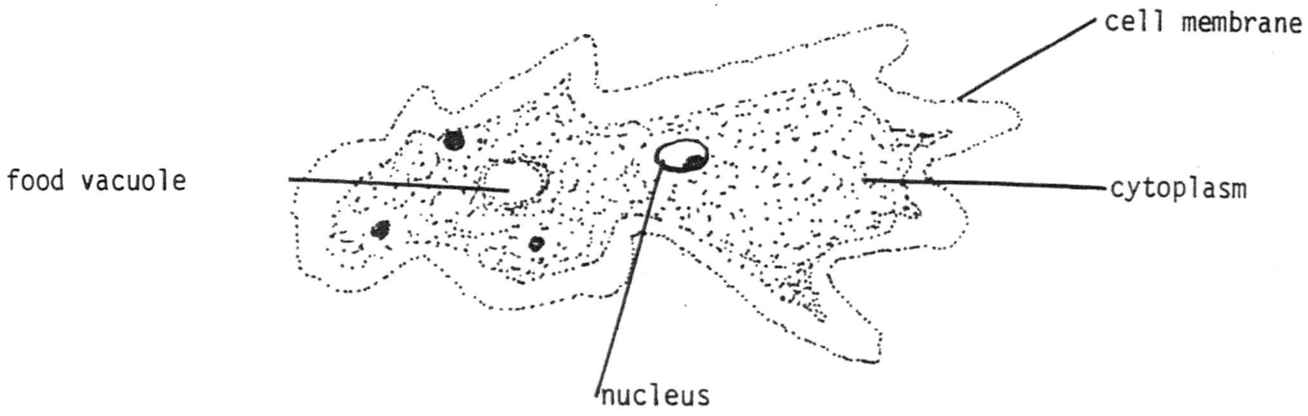

FORAMINIFERIDA SHELL TYPES

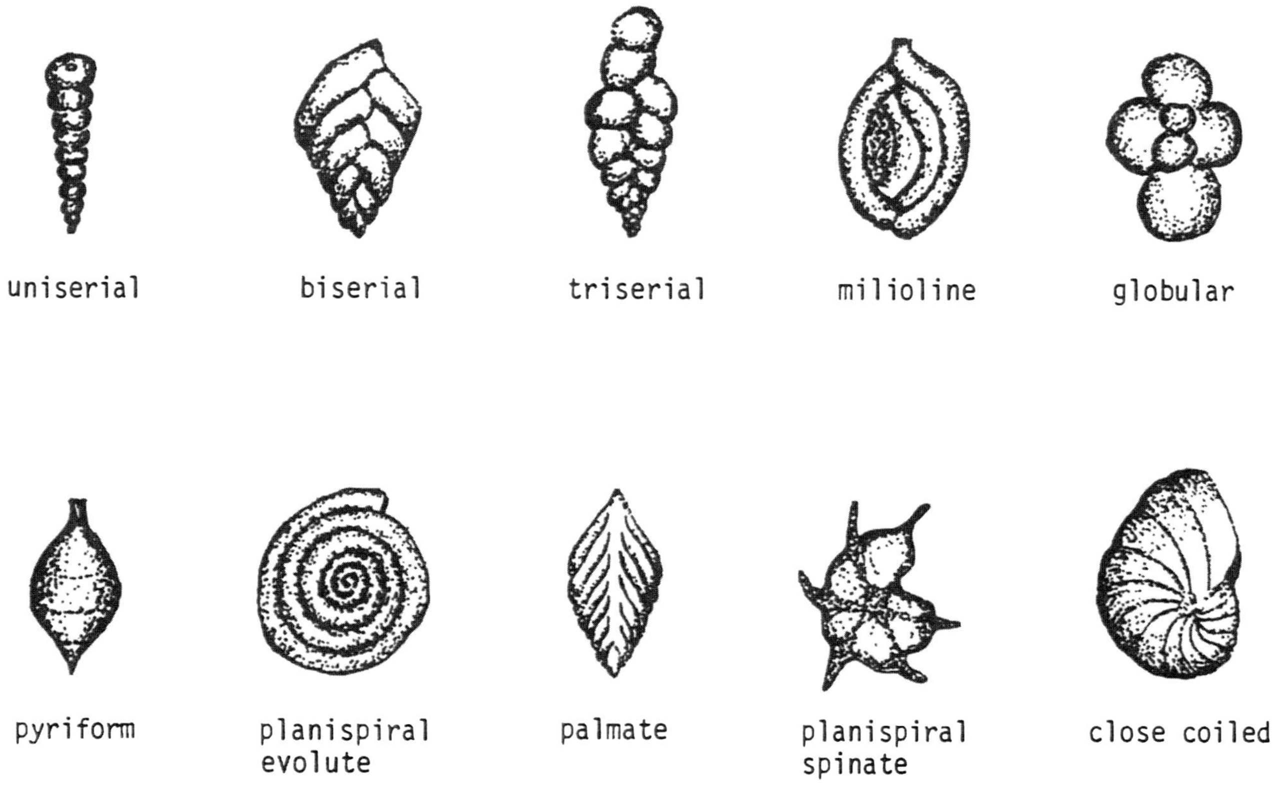

FORAMINIFERIDA MORPHOLOGY

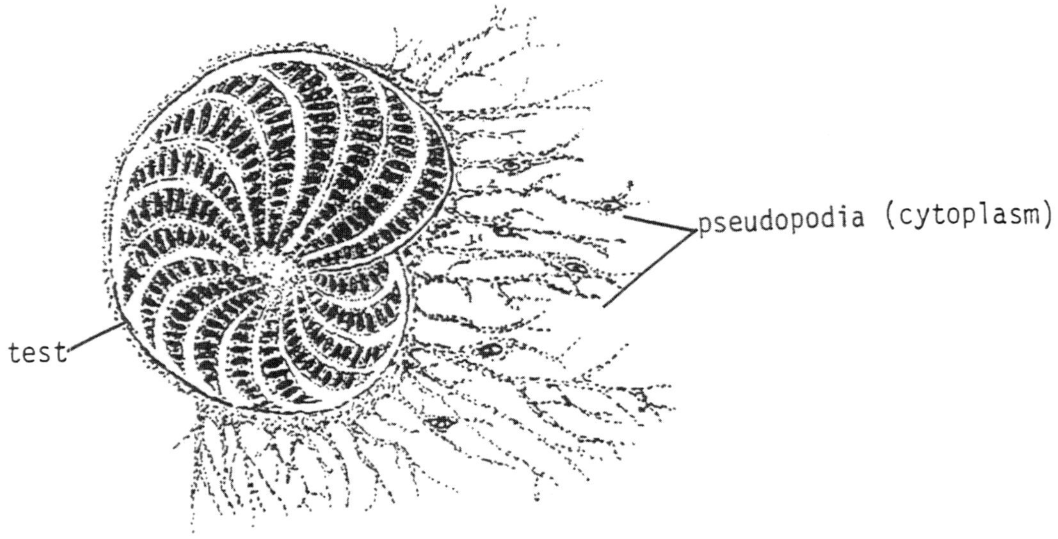

TYPES OF TESTS

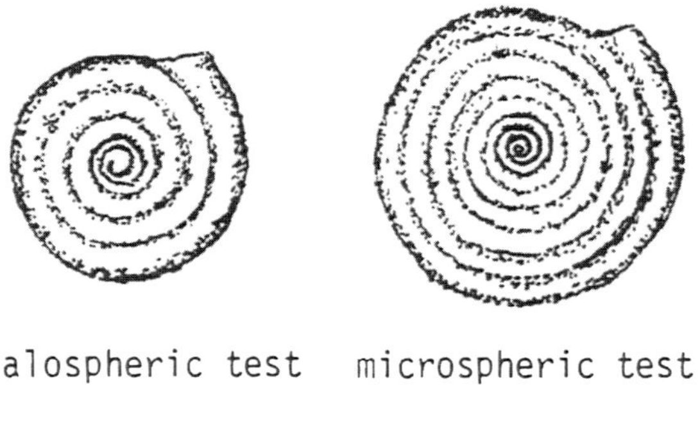

TEXTULARIINA

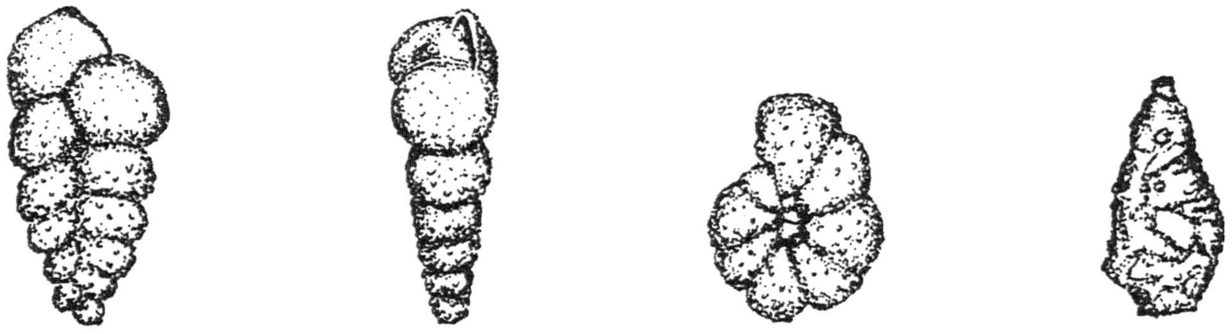

FUSULININA

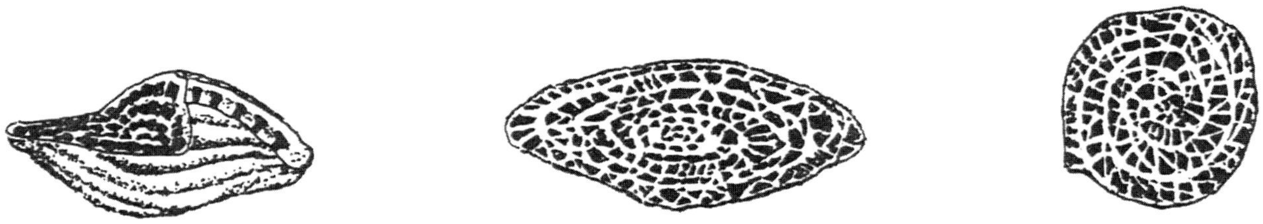

MILIOLINA

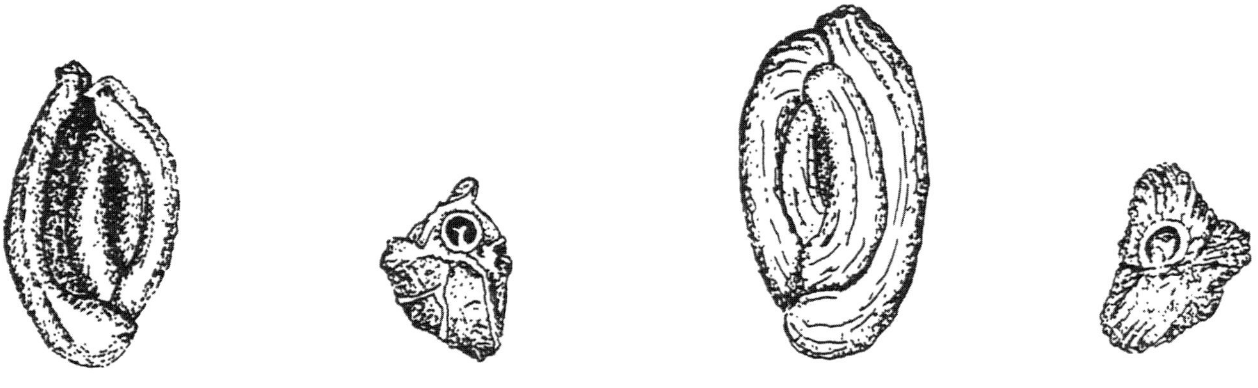

ROTALIINA
BENTHONIC FORMS

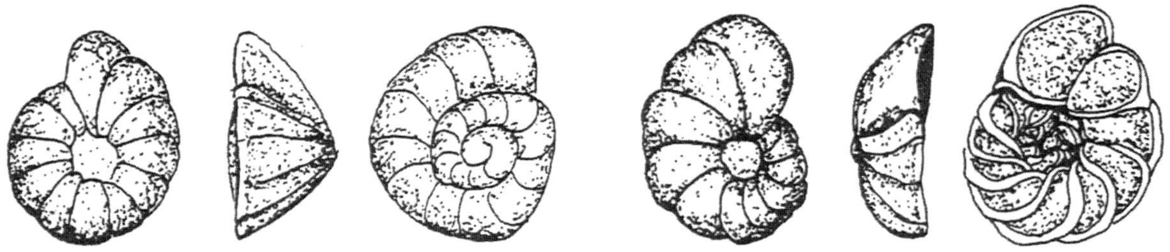

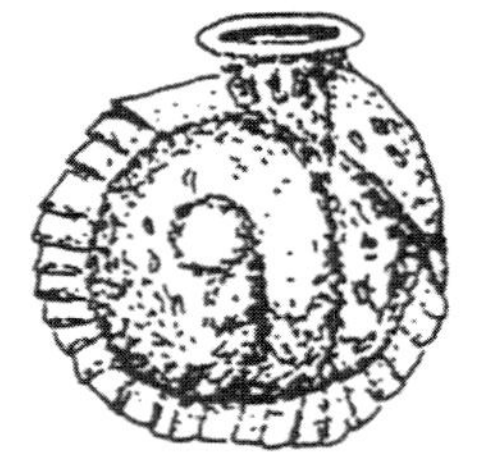

ROTALIINA
PLANKTONIC FORMS

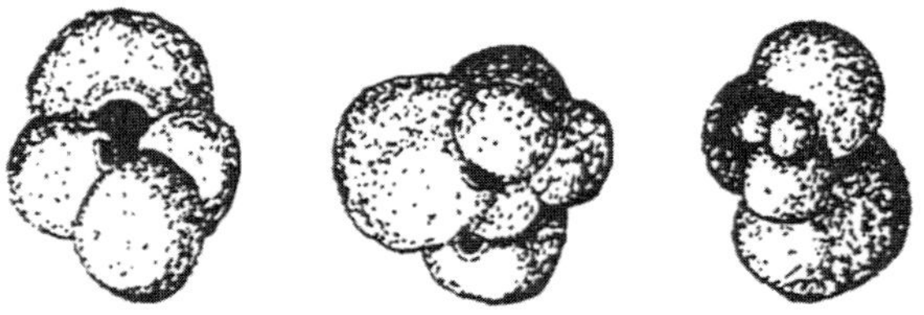

RADIOLARIAN MORPHOLOGY

RADIOLARIA

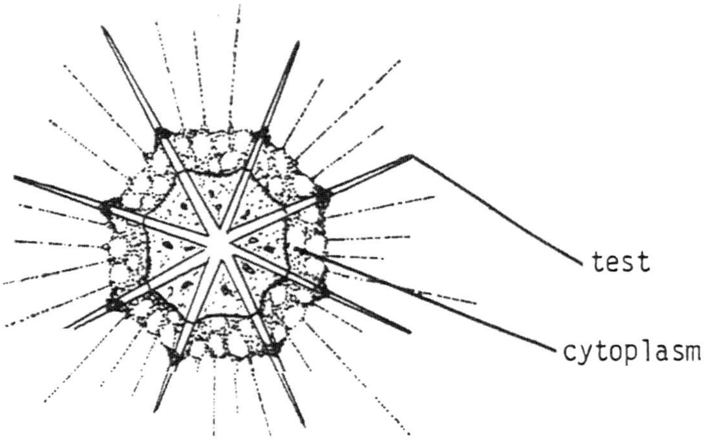

TYPES OF TESTS

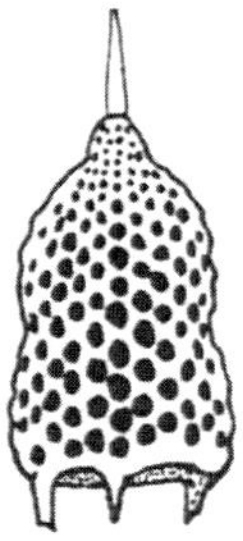
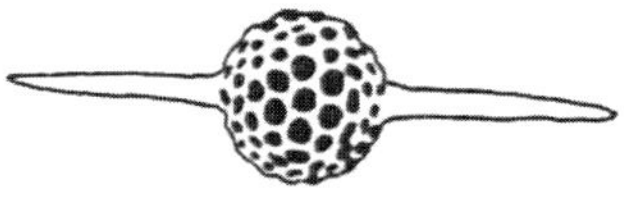
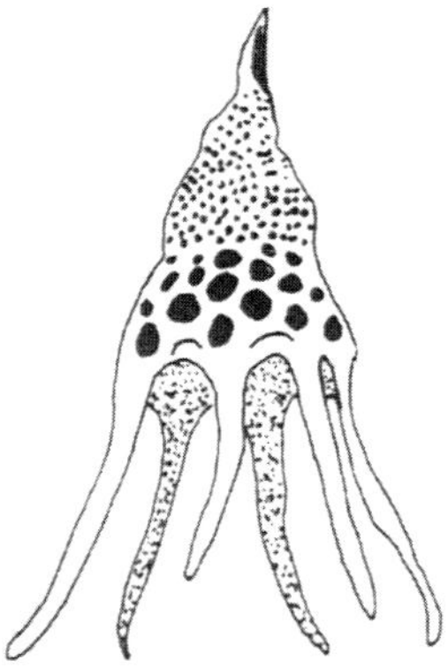

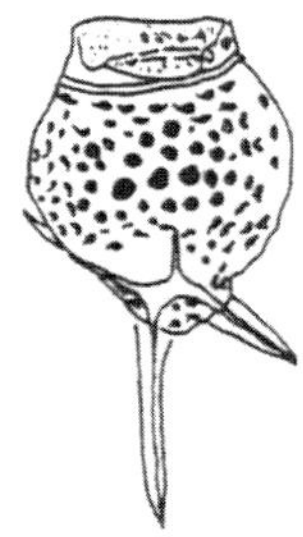
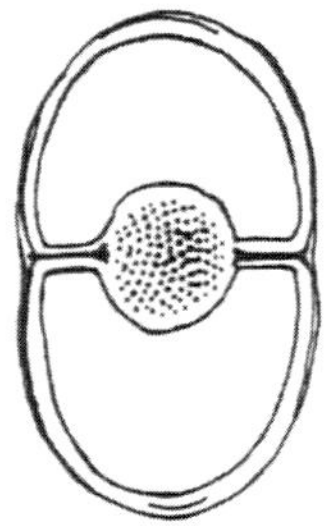
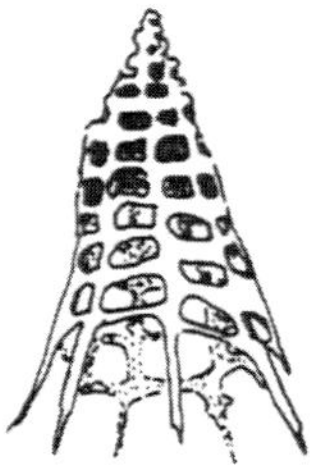
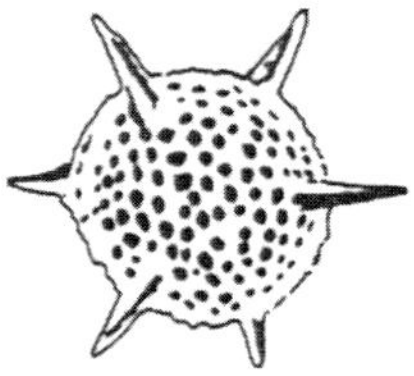
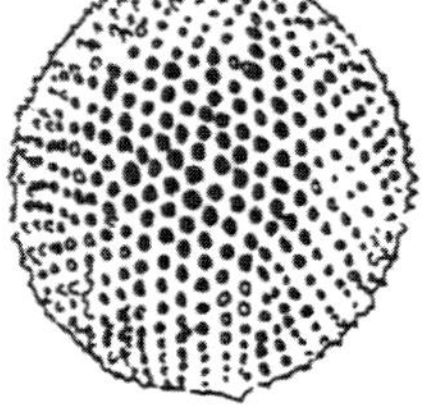

DIATOMS

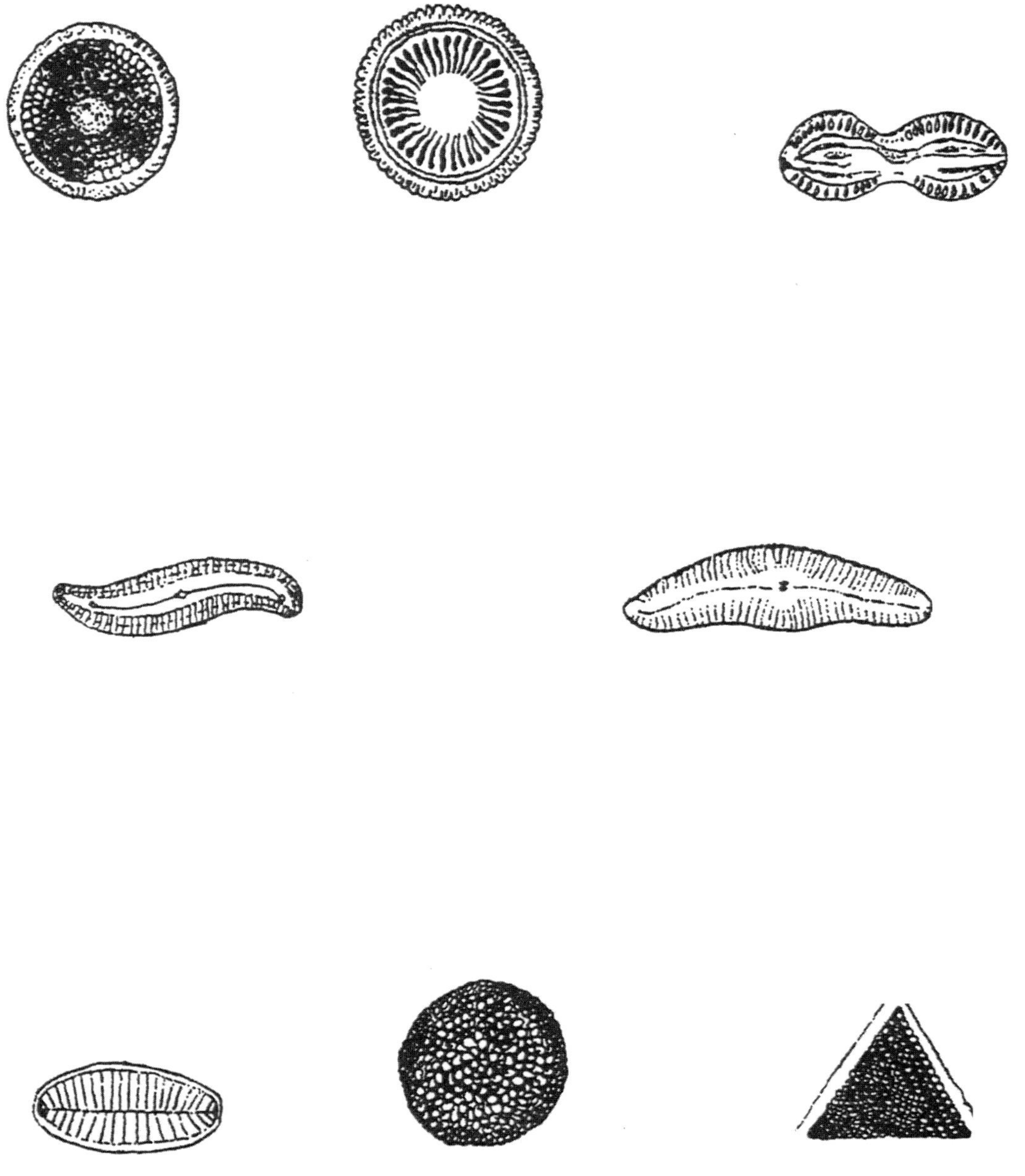

NOTES

Chapter 4: PORIFERA

A. Discussion

Sponges (Phylum Porifera) are the most primitive of the multicellular animals. They do not have true organs nor do they have true tissues. The body consists of a loose arrangement of three layers of cells. The epidermal layer has flattened polygonal cells. The middle layer, mesenchyme, consists of amoebocytes and skeletal material. The inner layer of cells, called collar cells or choanocytes, have flagella and line a central cavity, or spongocoel, and its canal-like pores.

Sponges have many shapes including both encrusting and upright forms. An idealized shape for study purposes is vase-like with an opening at the top called an osculum. Smaller openings, called pores, occur around the sides of the sponge. They allow water to pass through incurrent and radial canals from the outside to the central spongocoel.

As water passes through, food and oxygen are extracted with the aid of amoebocytes and collar cells. Waste products of metabolism are carried with the water out the osculum. The collar cells help move the water through more quickly by propelling it with their beating flagella.

Stiffness in the Porifera is achieved by spongin and/or spicules which make up the skeletal material. The spicules may be connected to form a rigid framework, in which case they are called megascleres, or they may be embedded in the flesh as microscleres. The spicules are composed of calcium carbonate or silicon dioxide and have a wide variety of shapes. Some of the more important spicule shapes are monaxons, triaxons, and tetraxons. The classification of sponges is in part based on the spicules' chemical composition and shape. Class Hexatinellida have siliceous triaxon spicules. Class Calcarea have calcareous spicules which are often of the monaxon type.

Sponges are typically sessile and live attached to the ocean floor or other substrate. They are distributed worlwide at many depths and in all seas. There are even a few fresh water species. Sponges with calcareous spicules generally live in water that is less than 200 meters deep. Sponges with siliceous spicules prefer cooler deep waters of more than 200 meters.

In the fossil record, sponges are usually represented by their spicules. If the spicules are connected, the shape of the once living sponge can be deduced. Unfortunately, many spicules in the fossil record are found disarticulated and the shape of the original sponge can only be imagined. Comparisons of fossil spicules with spicules from living specimens can sometimes help the paleontologist decide on a feasible shape. Sponges have a long history and extend from the Precambrian to the present.

B. Synopsis

Phylum Porifera [Cambrian-Recent]

- No true tissues or organs, cells relatively independent
- Mostly marine, sessile
- Skeleton consists of spongin or spicules which are either calcium carbonate or silicon dioxide

C. Representative Genera

Brachiospongia

Ordovician-Silurian

Astraeospongium

Silurian-Devonian

Astylospongia

Ordovician-Silurian

Palaeomanon

Silurian

Titusvillia

Carboniferous

Hydnoceras

Devonian-Mississippian

PORIFERA MORPHOLOGY

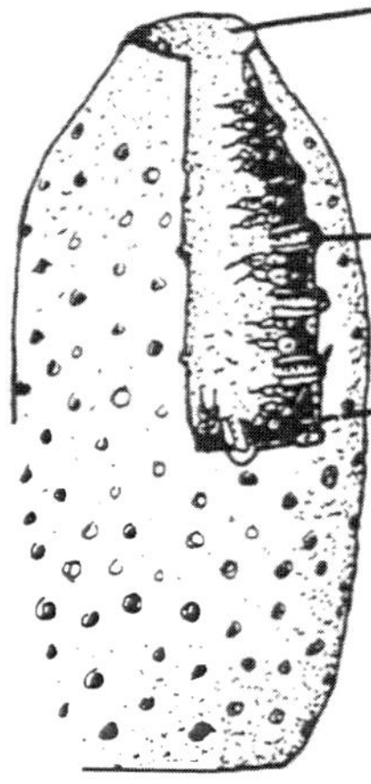

PORIFERA TYPES

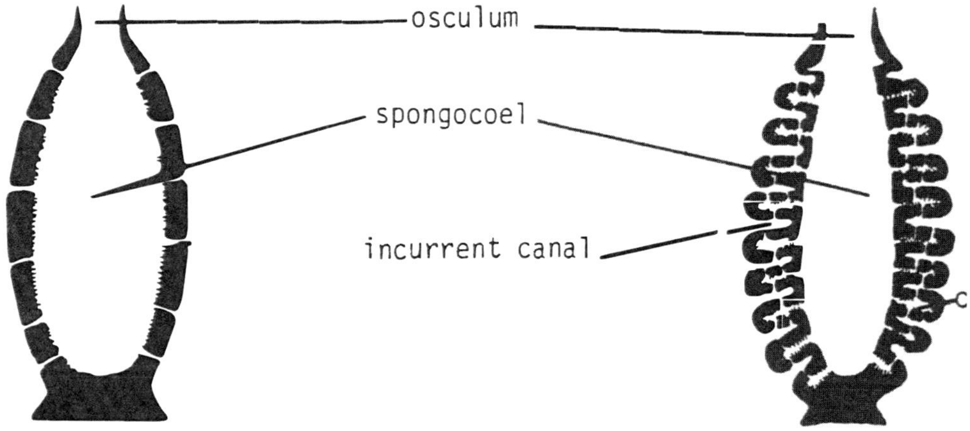

ascon body type

sycon body type

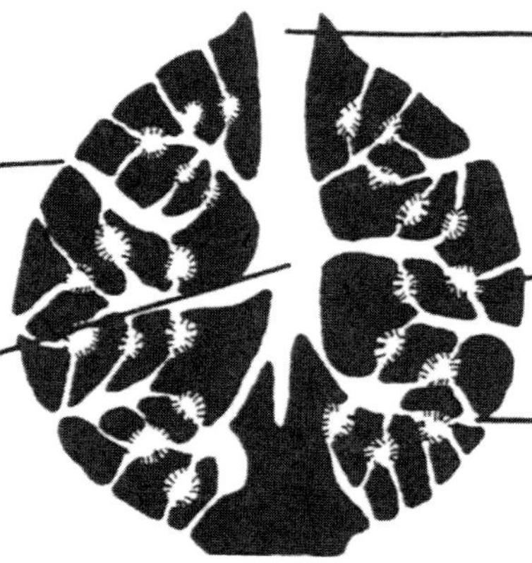

leucon body type

PORIFERA SPICULES

monaxon

triaxon

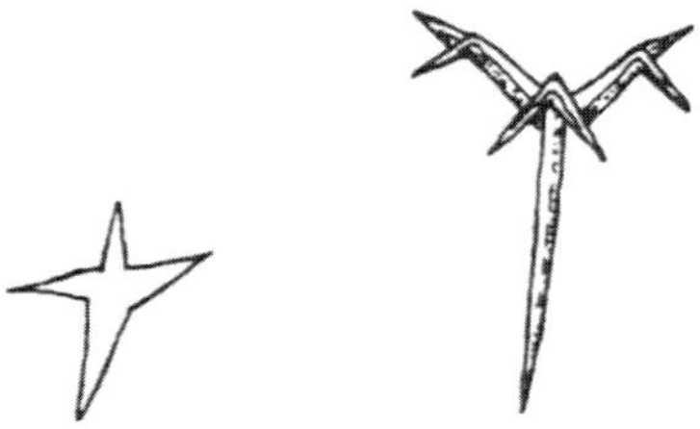

tetraxon

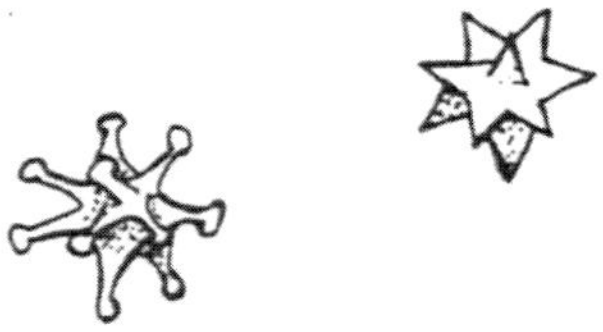

polyaxon

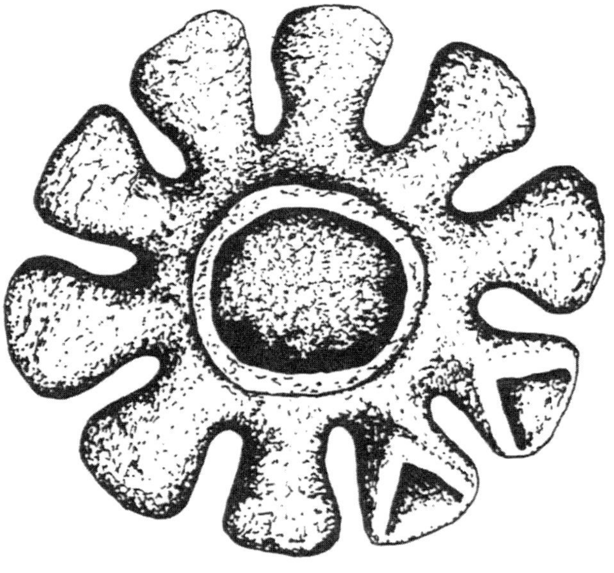

Brachiospongia

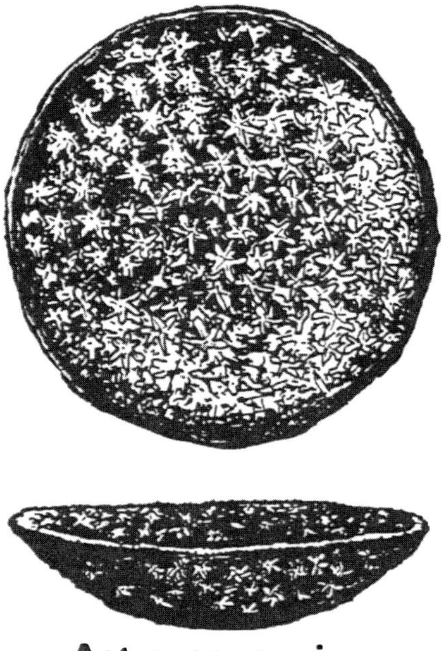

Astraeospongium

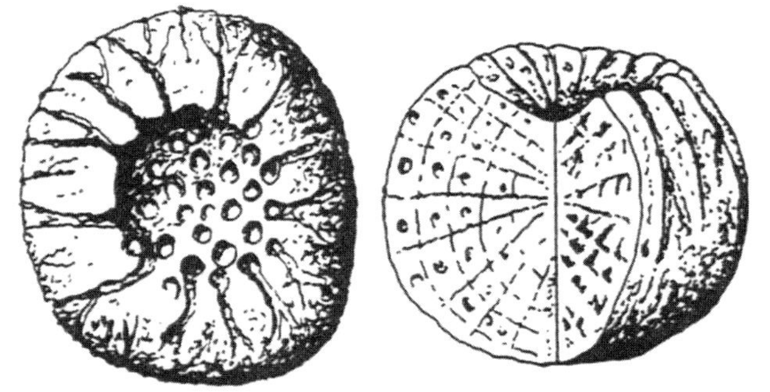

Astylospongia

Palaeomanon

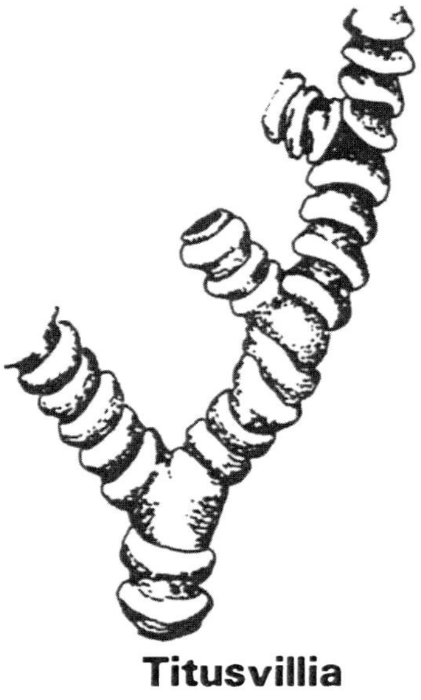

Titusvillia

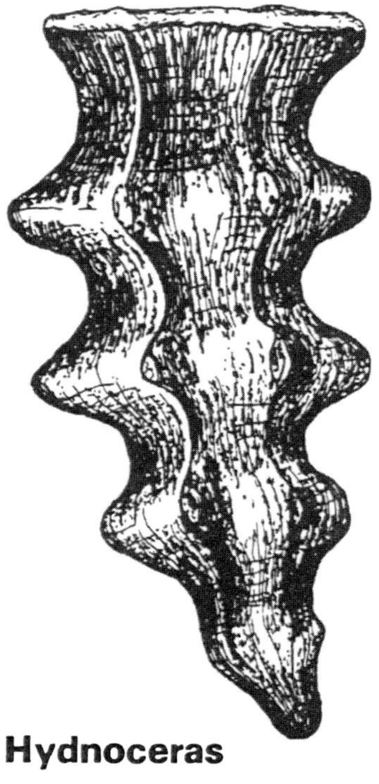

Hydnoceras

NOTES

Chapter 5: ARCHAEOCYATHA

A. Discussion

Phylum Archaeocyatha is a poorly understood group of organisms known by their peculiar calcareous fossils. They are found only in Cambrian formations, having become extinct before the Ordovician period. At various times, the archaeocyathids have been classified in the same phylum as either sponges or coelenterates, because they share characteristics common to each phylum. They are now generally considered a separate and distinct group.

Archaeocyathid fossils are generally cup or cone shaped. They have a central body cavity surrounded by a double set of walls. The outer and inner skeletal walls are joined by vertical septa or, occasionally, horizontal septa called tabulae. The space between the two walls is called the intervallum. The entire structure is perforated by small openings or pores interconnecting all of the skeletal spaces. Unfortunately there are no fossils which have preserved the soft anatomy and so its relationship to the skeleton is unknown.

In Cambrian paleocommunities, Archaeocyatha lived singly or formed colonies of sessile attached organisms in warm shallow seas. Because of their restricted geographical locality and importance as reef builders in equatorial seas, paleontologists have made use of the archeocyathids to determine the location of the Cambrian paleoequator.

B. Synopsis

Phylum Archaeocyatha [Cambrian]

Inner and outer walls, septa, occasionally tabulae

Calcium carbonate skeleton

Bethonic in shallow warm seas

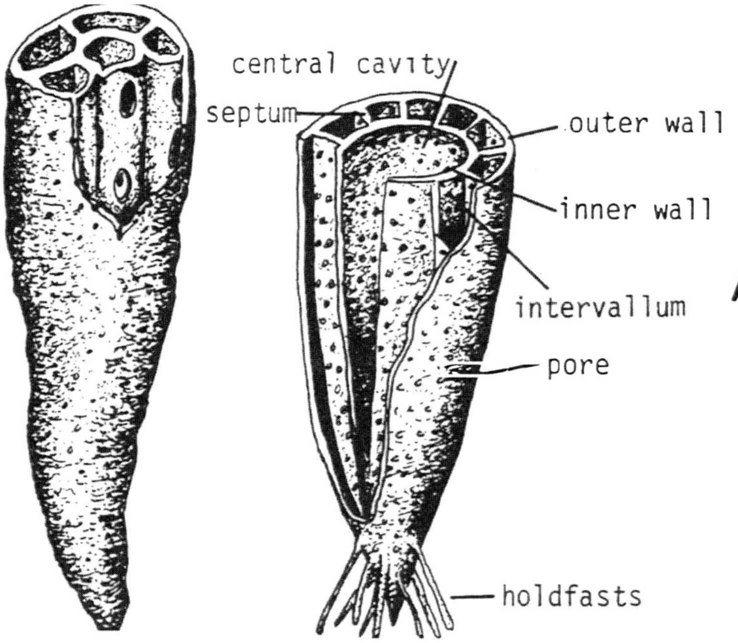

ARCHAEOCYATHID MORPHOLOGY

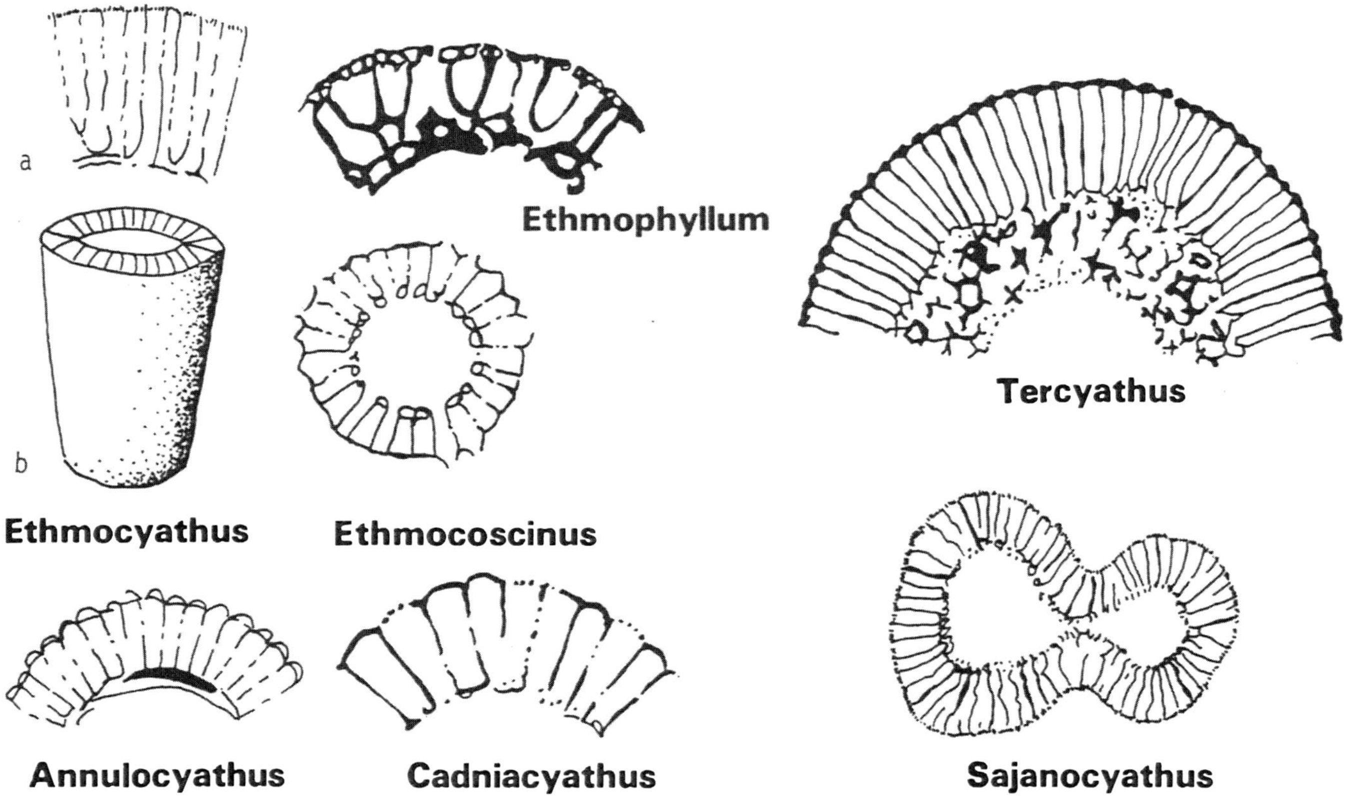

CROSS SECTIONS SHOWING SEPTA OF DIFFERENT ARCHAEOCYATHIDS

(After Treatise)

ARCHAEOCYATHID BODY TYPES

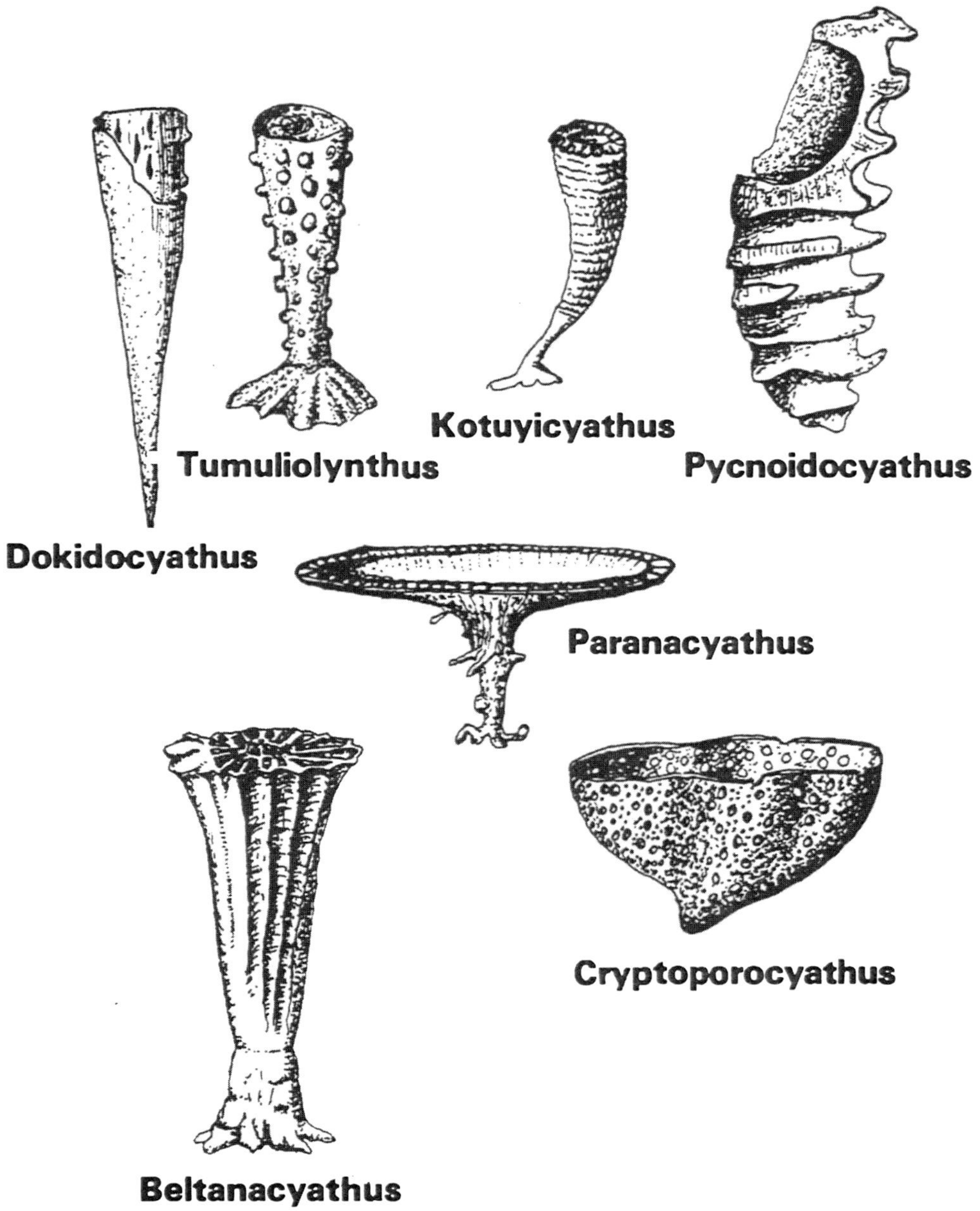

(After Treatise)

NOTES

Chapter 6: RECEPTACULITIDS

A. Discussion

One of the least understood groups in the fossil record is the receptaculitids, a group of organisms whose phyletic affinities were until very recently poorly understood. Paleontologists have at various times classified them as sponges, coelenterates, and echinoderms. Although current opinion favors placing the receptaculitids with calcareous algae, we include them here because they are so often found associated with invertebrate assemblages.

Receptaculitids inhabited marine waters and were probably benthonic. The shape of the fossils may be globular, bowl, or disk-like. The disk forms are either slightly rounded or flat with upturned edges much like a saucer. Occasionally a central body cavity is found, especially in the globular forms, and some paleontologists believe that bowl and disk shapes merely represent examples of collapsed globular forms or the upper portion of their skeletons.

The skeleton of a receptaculitid consists of numerous parallel columns or laterals. The ends of each lateral extend radially, forming lateral projections which meet similar projections on adjacent laterals. These lateral surfaces appear as facets on the exterior of the receptaculitid fossil, creating readily identifiable whorls or spirals radiating from the apex of the organism much like the arrangement of seeds in a sunflower.

Receptaculitids lived from the Ordovician to the Devonian period. Their distribution was widespread; and they were particularly abundant in the central and western parts of the United States in the middle Ordovician period.

B. Synopsis

Kingdom Plantae

Phylum Chlorophyta

Receptaculitids

Ordovician-Devonian

Calcium carbonate skeleton, sunflower appearance

RECEPTACULITIDS

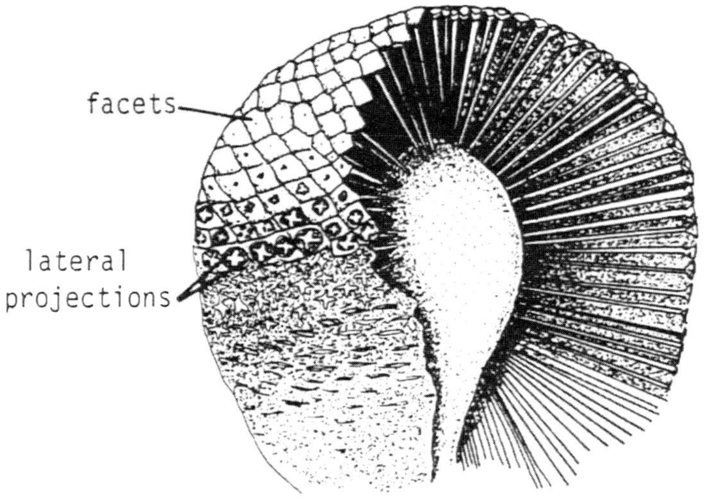

(After Nitecki)

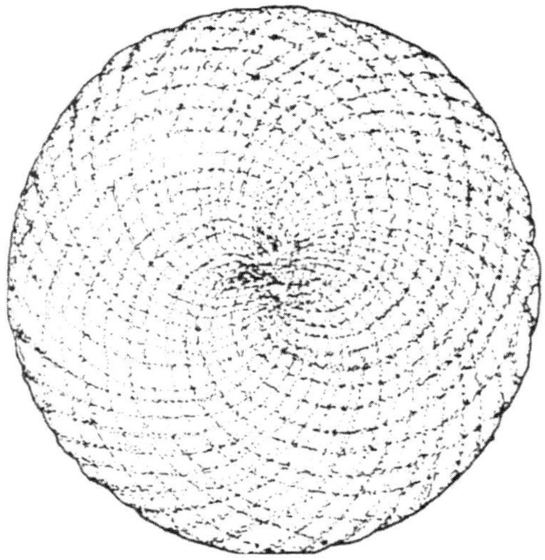

surface view

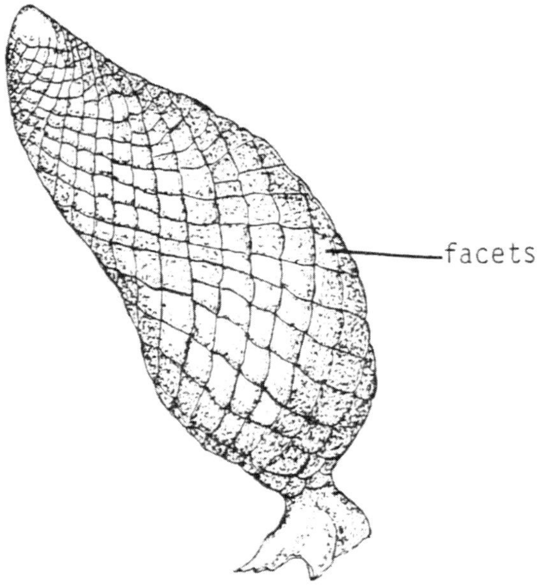

(After Nitecki)

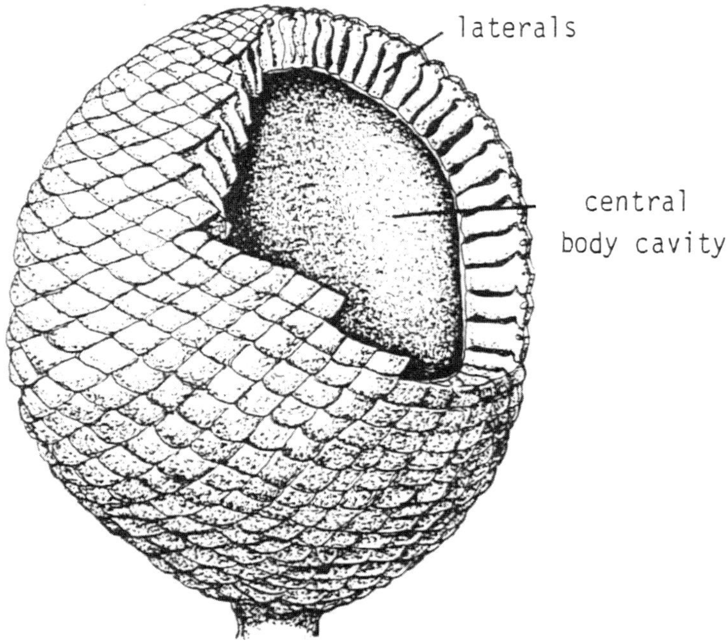

(After Nitecki)

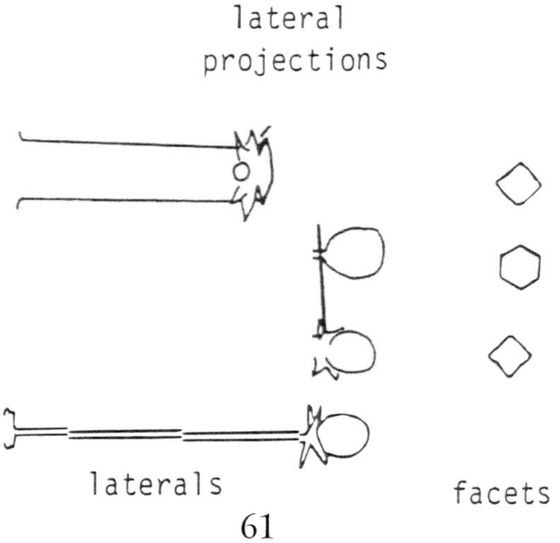

NOTES

Chapter 7: COELENTERATA

A. Discussion

Coelenterates (Phylum Coelenterata) are familiar inhabitants of tidepools, open oceans, and ocean floors. They include among their diverse members sea anemones, jellyfish, corals, sea pens, sea fans, and hydras.

A coelenterate has three body layers: an outside layer or epidermis, a middle layer called the mesoglea, and an inner layer or gastrodermis, which is a layer of cells lining the gastrovascular cavity. There are no special organs for respiration or excretion and the nervous system is very primitive.

The body of a coelenterate is radially symmetrical and resembles a sack. A single opening serves as both mouth, for ingestion of prey, and anus, for the elimination of waste products. The inside of the sack is a gastrovascular cavity for the digestion of food. Surrounding the mouth are tentacles which contain stinging cells or nematocysts. The tentacles capture and hold prey while the nematocysts are discharged into the victim, immobilizing it. Two different forms of the body exist at different stages of the life cycle. One is the polyp stage during which the organism has a stalk and is attached to the substrate. The mouth and tentacles extend upward. The second or medusa state is somewhat different in that the organism is not attached but floats or swims with the mouth and tentacles extending downward. A single species within the phylum can have either one or both polyp and medusa stages.

Reproduction in the coelenterates can be either sexual or asexual. During the asexual phase, reproduction occurs by budding from a polyp. A small outgrowth containing some of the gastrovascular cavity forms on the animal's side. After a mouth and tentacles develop, the bud detaches from the parent and becomes a free swimming medusa. The medusa is the beginning of the sexual phase of reproduction. Eggs and sperm develop in the gonads and are released into the water where the gametes unite to form a larva. After a short free swimming existence, the larva attaches to a hard surface and develops into a polyp. The reproductive mechanism described above is, of course, idealized and many variations on the pattern exist.

There are three important classes of coelenterates: Hydrozoa, Scyphozoa, and Anthozoa. The Hydrozoa have either or both the polyp and medusa states in their life cycles. Scyphozoa, which include jellyfish, have a conspicuous medusa stage and inconspicuous polyp stage. Anthozoa have completely lost the medusa stage and have a prominent polyp stage.

Hydrozoans are represented in the fossil record by the stromatoporoids, although their inclusion in this class is somewhat debatable. The stromatoporoids were important companions to rugose and tabulate corals as reef builders during the Silurian and Devonian periods. Close observation of fossil stromatoporoids in cross section discloses very thin layers or laminae, a diagnostic characteristic of the group. On the surface the colonies are sheet-like or denroid masses.

Scyphozoa are not well represented in the fossil record because of the absence of hard preservable parts. Usually impressions are the only record we find of their existence. One group, however, did have a chitinophosphate covering which was occasionally fossilized. These were the Conularia, now extinct, whose body shape resembles an inverted pyramid showing many prominent horizontal ridges. Because in a few rare specimens tentacles are evident, the Conularia have been assigned to the class Scyphozoa.

Of all the fossil coelenterates, corals are the best known and the most abundant because they produce calcareous exoskeletons. The hard calcium carbonate parts of corals are more readily preserved than the soft tissues of the other groups of coelenterates.

Corals are divided into several orders. The important Paleozoic order Rugosa contains coral fossils which look like wrinkled horns. They have vertical septa which radiate inward from the sides of the exoskeleton and are an important characteristic that separates them from the order Tabulata, another important Paleozoic group. Tabulate corals have conspicuous horizontal septa called tabulae and relatively inconspicuous septa. The two orders also differ in that rugose corals have both solitary and colonial species, whereas tabulate corals are all colonial. Both orders were important reef builders in the middle and late Paleozoic communities. Because of their generally short stratigraphic ranges from the Ordovician to the Permian period, species of these two coral orders are important index fossils.

The order Scleractinia contains modern corals. They are represented in the fossil record only since the Triassic period. A few scleractinian corals superficially resemble rugose corals; however, a careful examination of the septa will help distinguish them. Rugose corals have septa displaying four fold symmetry, hence the name frequently applied to them of tetracoral. Scleractinian corals or hexacorals have septa which show six fold symmetry. As a general rule, the septa of scleractinian corals look more complex than those of rugose corals and have a white bleached appearance which is quite unlike rugose corals.

In the marine environment, modern corals superseded rugose and tabulate corals as reef builders and have been very important in that role since the Mesozoic.

B. Synopsis

Phylum Coelenterata [Precambrian-Recent]

- Radial symmetry
- Three layers of cells: epidermis, mesoglea, gastrodermis
- Mouth and stomach, no anus
- Medusa and polyp stages
- Class Hydrozoa [Precambrian-Recent]
  - Polyp, medusa or both stages
- Class Scyphozoa [Precambrian-Recent]
  - Medusa stage prominent
- Class Anthozoa [Ordovician-Recent]
  - Polyp stage only
  - Calcium carbonate exoskeleton in coral members
  - Order Rugosa (tetracorals) [Ordovician-Permian]
    - Septa conspicuous
  - Order Tabulata [Ordovician-Permian]
    - Tabula conspicuous
  - Order Scleractinia (hexacorals) [Triassic-Recent]
    - Modern corals

Chapter 7: Coelenterata

C. Representative Genera

Stromatopora (Hydrozoa)

Ordovician-Permian

Conularia (Scyphozoa)

Cambrian-Permian

Heliophyllum (Rugosa)

Devonian

Lambeophyllum (Rugosa)

Ordovician

Hexagonaria (Rugosa)

Devonian

Favosites (Tabulata)

Ordovician-Devonian

Honeycomb coral

Halysites (Tabulata)

Ordovician-Silurian

Chain coral

Parasmilia (Scleractinia)

Cretaceous-Recent

Montlivaltia (Scleractinia)

Triassic-Cretaceous

Diploria (Scleractinia)

Cretaceous-Recent

Fungia (Scleractinia)

Miocene-Recent

COELENTERATE MORPHOLOGY

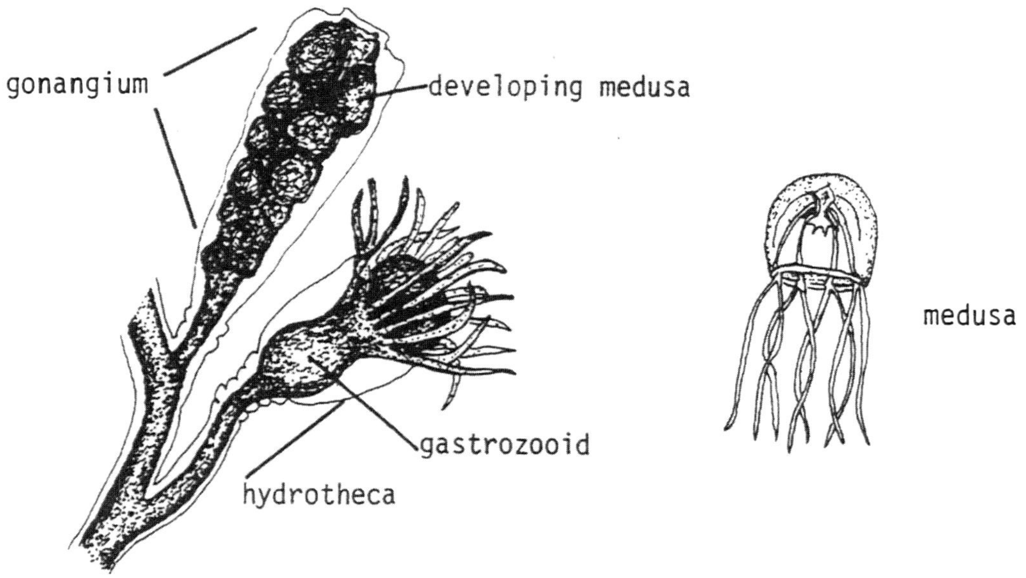

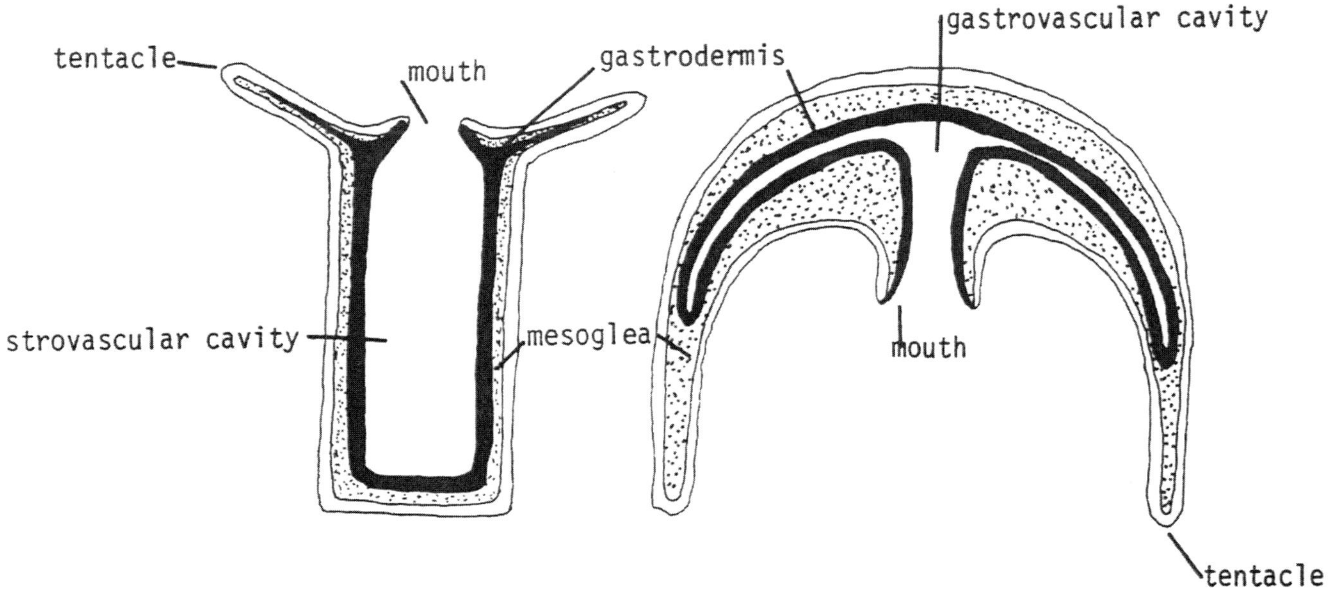

polypoid body form

medusoid body form

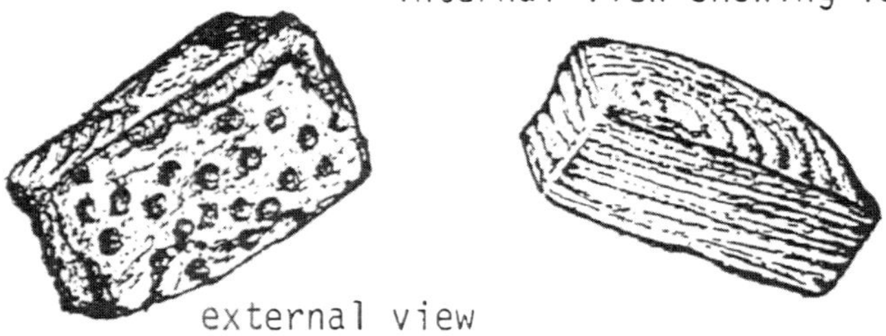

Stromatopora

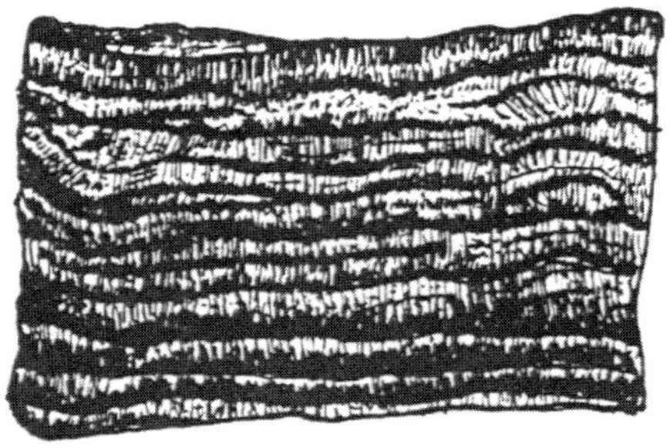

Actinostroma,
a stromatoporoid

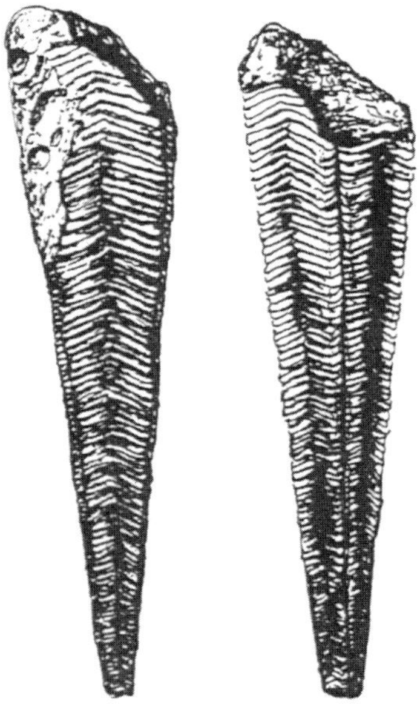

Conularia

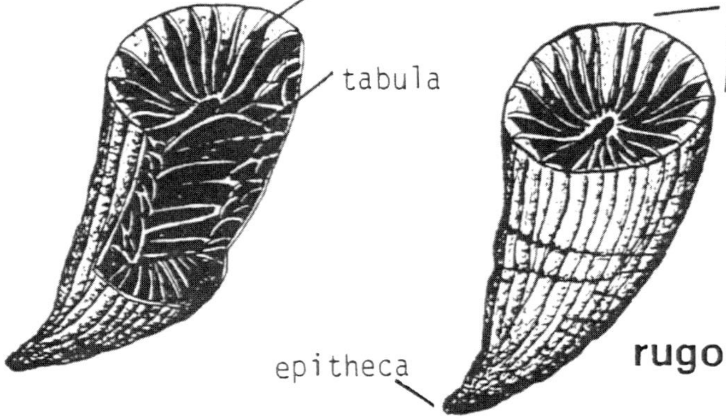

rugose coral

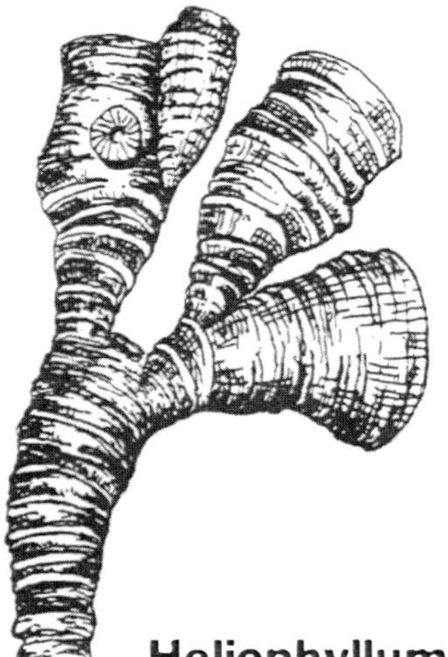

Heliophyllum

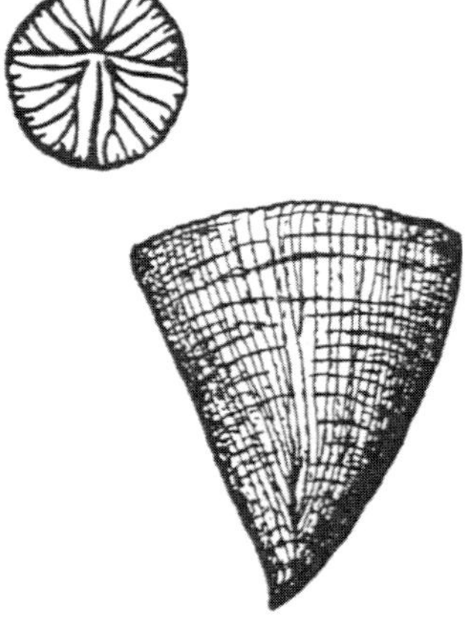

Lambeophyllum

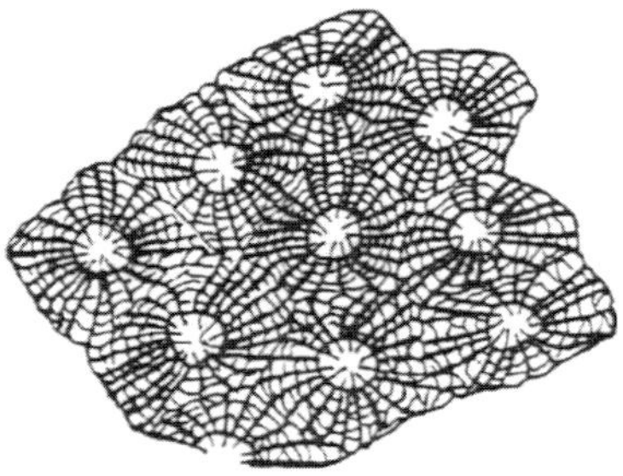

Hexagonaria

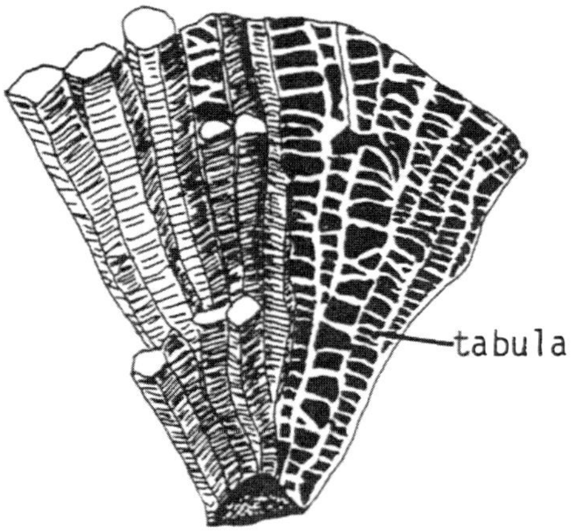

Favosites

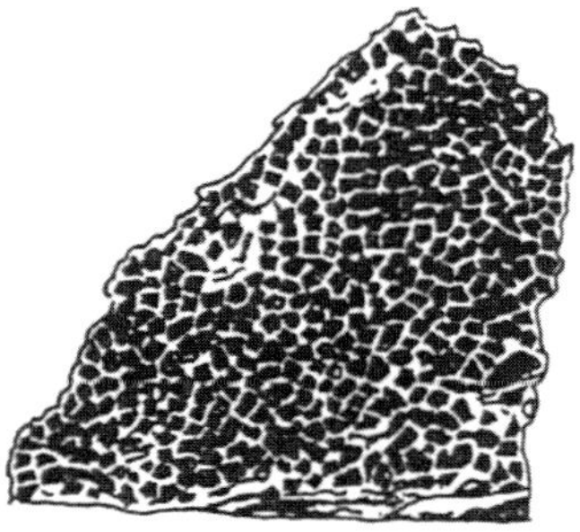

Favosites

Halysites

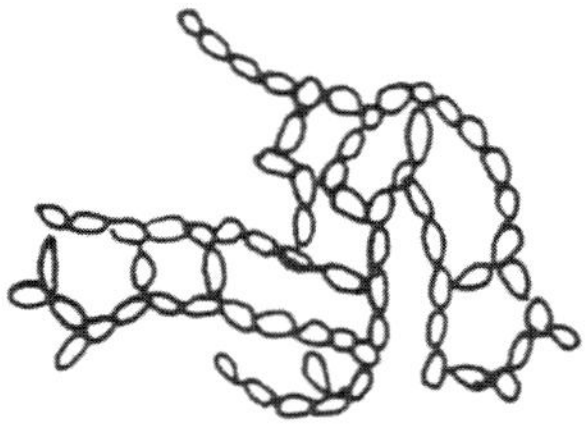

Halysites

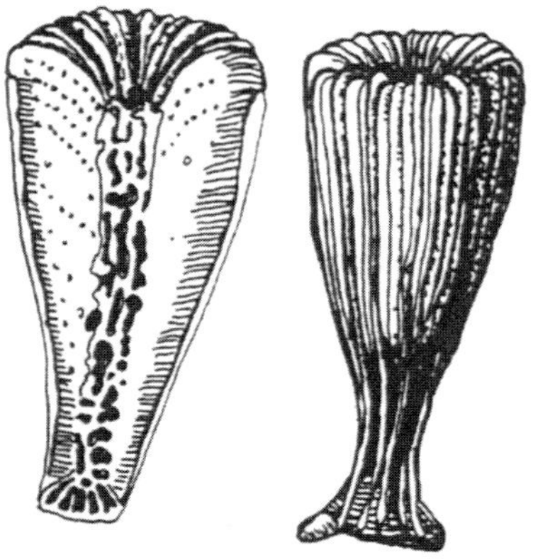

Parasmilia

(After Treatise)

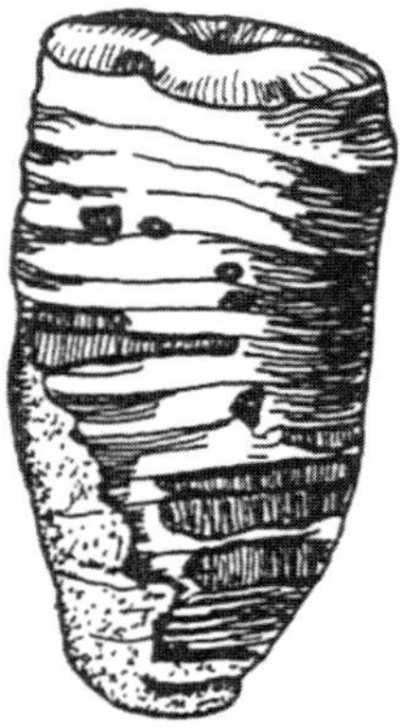

Montlivaltia

(After Treatise)

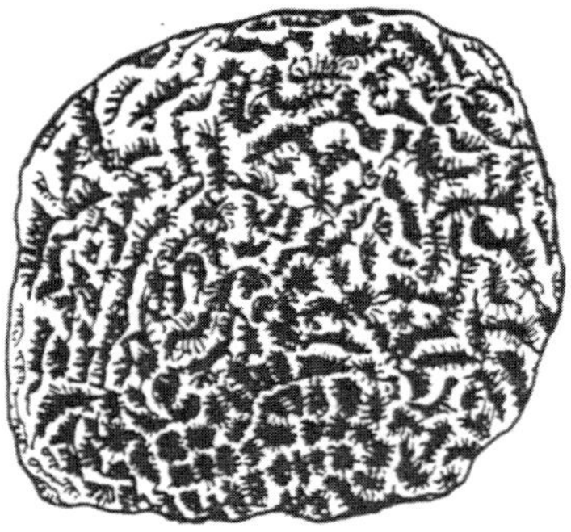

Diploria

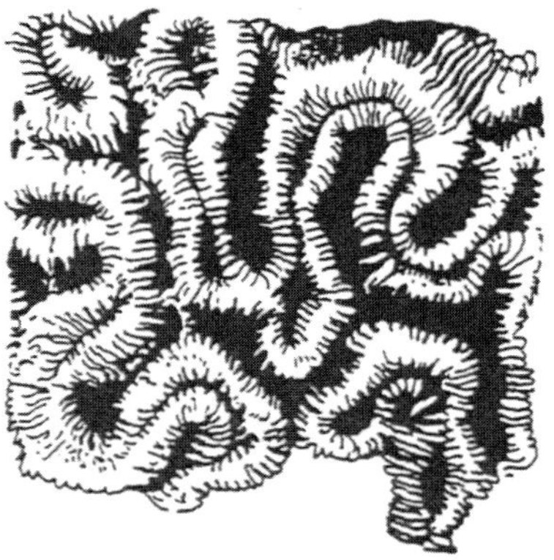

Diploria, enlarged

(After Treatise)

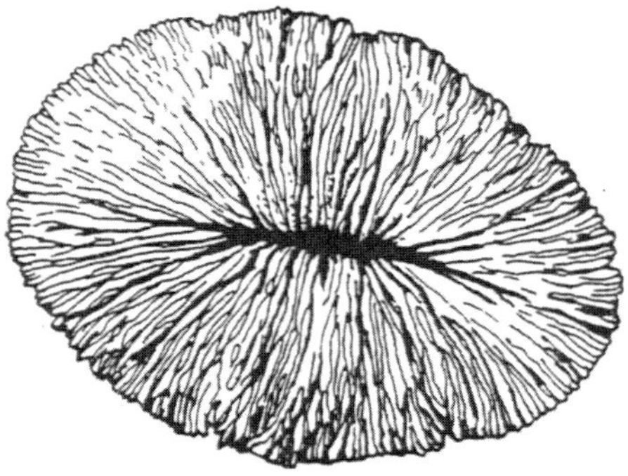

Fungia

NOTES

Chapter 8:
BRYOZOA
(Ectoprocta or Polyzoa)

A. Discussion

Although Bryozoa are placed in their own phylum, they share some unique characteristics with the phyla Brachiopoda and Phoronida. The members of these phyla are called lophophorates because they all possess a specialized food gathering organ called a lophophore. The lophophore consists of a circular or "U" shaped array of ciliated tentacles. The cilia propel water through the lophophore and small animals or plankton are entrapped and ingested, thus providing nourishment to the bryozoan, phoronid, or brachiopod. The lophophores also have in common a "U" shaped digestive system with both a mouth and an anus, indicating an increase in complexity beyond the coelenterate body plan. In addition, the larvae of the lophophorates show similar developmental stages which are unlike other larvae in the animal kingdom. Beyond these basic characteristics the Bryozoa, Phoronida, and Brachiopoda are quite different.

Bryozoa are all colonial organisms having no separate vascular, respiratory, or excretory organs. Individual bryozoa secrete a proteinaceous, chitinous, or calcareous exoskeleton which is called a zooecium. Many zooecia (plural of zooecium) make up an entire colony or zoarium. An individual zooecium may be shaped like a tube, vase, or flattened box. One side is perforated by a small opening through which the lophophore is extended into the water for feeding. The lophophore can also be retracted through the opening in case of danger. In many species a protective cover called an operculum is pulled over the lophophore opening.

Most Bryozoa are hermaphroditic, that is, an individual produces both sperm and eggs. The gametes are usually released at different times so that cross fertilization rather than self fertilization is the preferred mechanism for reproduction. The fertilized egg is brooded internally in some species and released directly into the water in others. After a short free swimming life, the developing larva settles onto a substrate and attaches. Asexual budding of the larva produces a bryozoan colony. The colony shape varies from flat encrusting forms to upright branching colonies. These shapes are related to the habitats of the Bryozoa; encrusting types live in high energy environments, and upright forms live in quieter waters.

Bryozoa are abundantly represented in the fossil record from Ordovician time to the present. Some orders had relatively short biostratigraphic ranges and are useful for dating and correlation. Others grew in specific environments, such as high or low current regimes and are excellent indicators of paleoenvironmental conditions.

B. Synopsis

Phylum Bryozoa (Ectoprocta) [Ordovician-Recent]

- Lophophore
- "U" shaped digestive system with mouth and anus
- Exoskeleton of protein, chitin, or calcium carbonate
- Sexual and asexual reproduction

Chapter 8: Bryozoa

C. Representative Orders

Order Ctenostomata

Ordovician-Recent

"Comb-mouth", zooecium composed of chitin or protein

Individuals develop by budding from a stolon

Order Cryptostomata

Ordovician-Permian

Tube shaped zooecium composed of calcium carbonate

Colonies are frond or ribbon shaped

Archimedes

Mississippian-Permian

Fenestrellina

Carboniferous-Permian

Order Cyclostomata

Ordovician-Recent

Zooecium is tube shaped, composed of calcium carbonate, no diaphragms

Order Trepostomata

Ordovician-Permian

Zooecium is tube shaped with diaphragms, composed of calcium carbonate

Stony bryozoans, zooarium massive, lamellar, or stem shaped

Order Cheilostomata

Jurassic-Recent

Zooecium is commonly shaped like flattened box, composed of chitin or calcium carbonate

Zoarium encrusting, frond, or stem shaped

BRYOZOA MORPHOLOGY

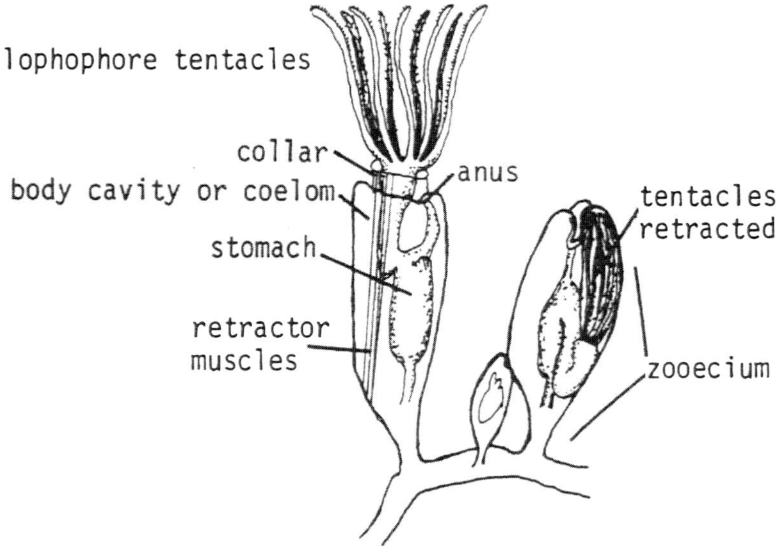

CTENOSTOMATA

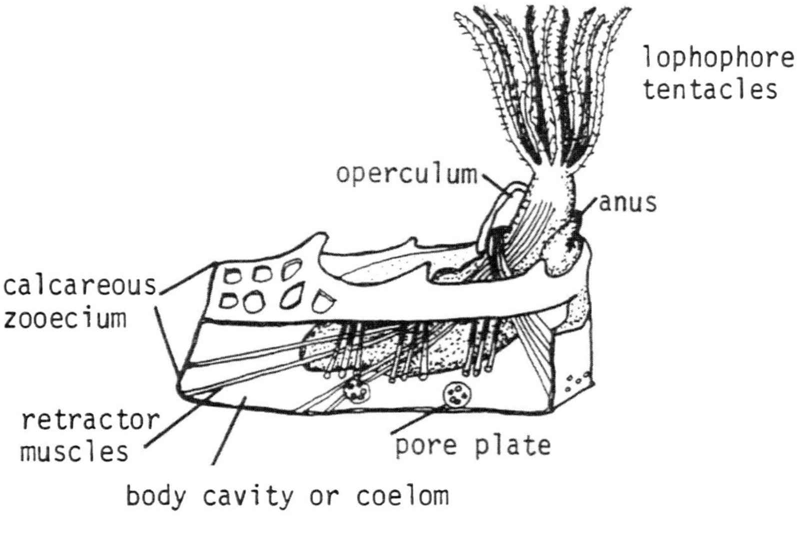

CHEILOSTOMATA

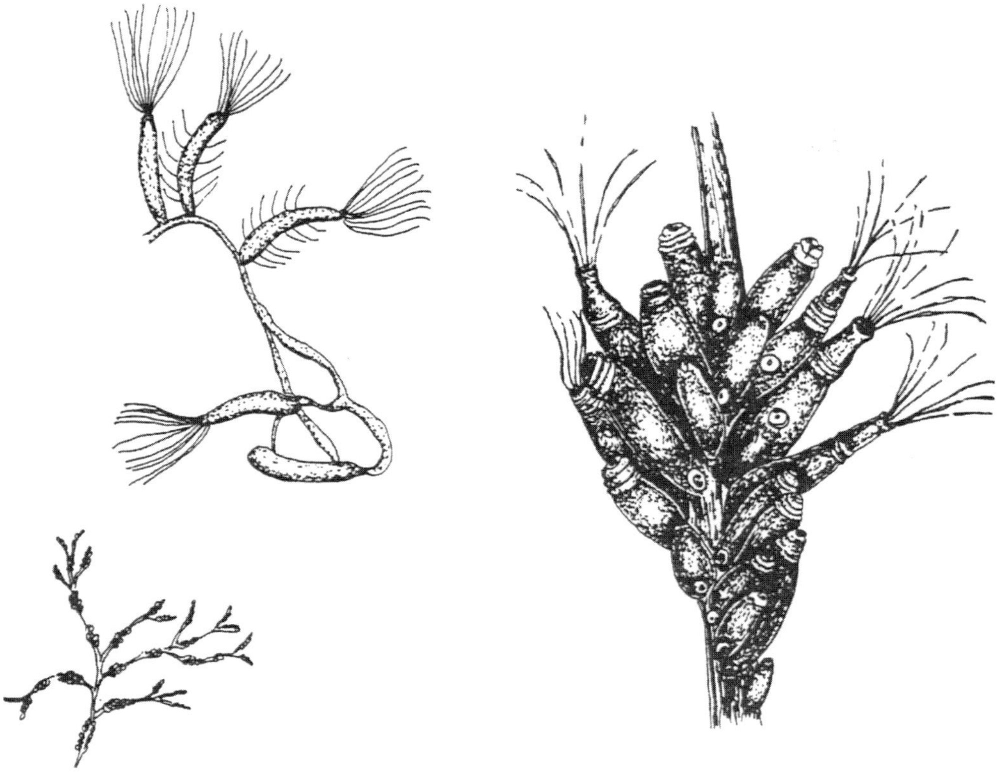

CTENOSTOMATA

CRYPTOSTOMATA

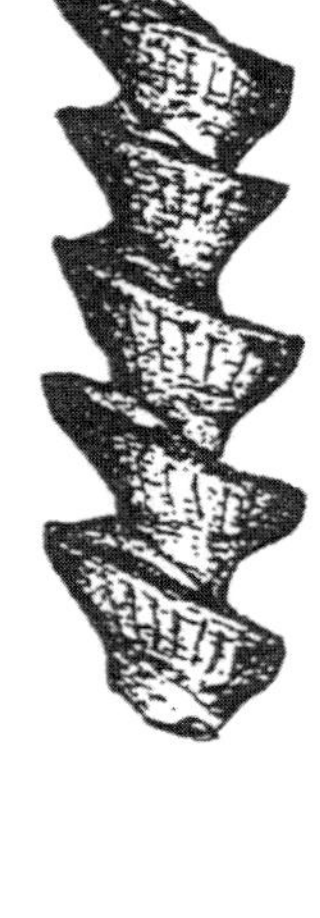

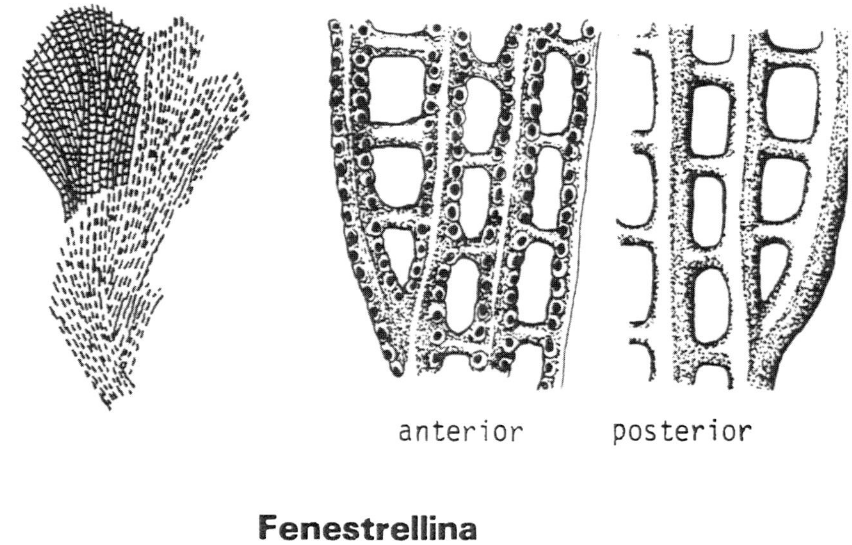

Archimedes **Fenestrellina**

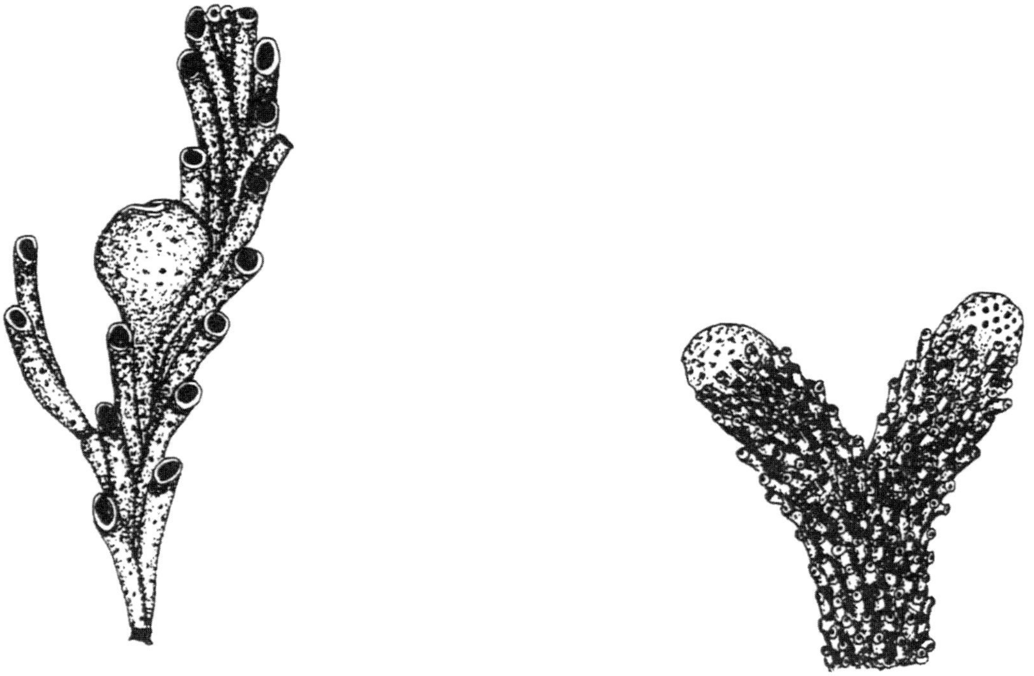

CYCLOSTOMATA

TREPOSTOMATA

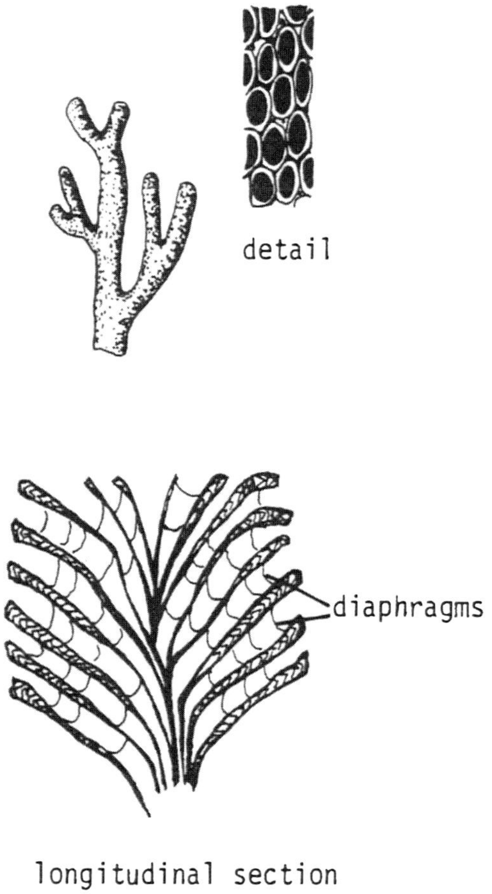

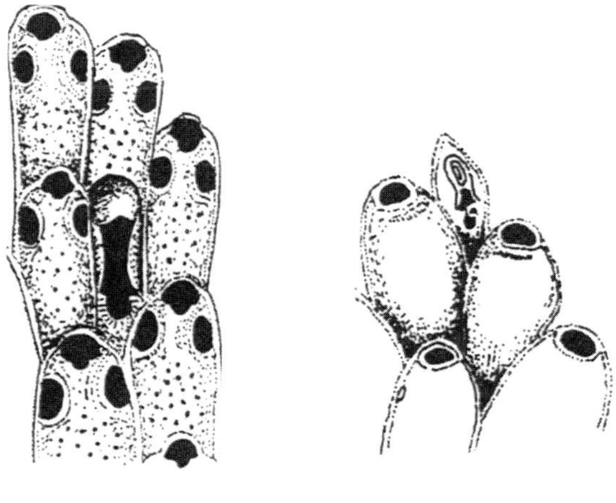

CHEILOSTOMATA

NOTES

Chapter 9: BRACHIOPODA

A. Discussion

The animals of the phylum Brachiopoda are similar to the Bryozoa discussed earlier. They possess a lophophore, a "U" shaped digestive system and a unique larva. However, unlike the Bryozoans, Brachiopods have developed a bivalved shell which is secreted by a mantle. The shells are responsible for Brachiopods having been classified as molluscs by early researchers. Actually the shells of brachiopods are quite unlike those of molluscs in that they enclose the soft body of the animal dorsoventrally. Molluscan shells, or valves, enclose the body laterally. In addition, the two valves of Brachiopods are unequal in size with the ventral valve usually larger than the dorsal valve. The valves of molluscs are symmetrical and of the same size and shape. Paleontologists prefer to not use the terms dorsal and ventral for the brachiopod valves and substitute pedicle valve for the ventral valve because often a pedicle or stalk extends from that valve to the substrate where it attaches. The dorsal valve is referred to as the brachial valve because the brachidium, a skeletal structure giving support to the lophophore, is located on that valve. The brachidium can often be identified in fossil specimens and in some animals is diagnostic for individual species.

Brachiopoda are more highly developed than their near relatives the Bryozoa. They have evolved a primitive heart and kidneys as well as a "liver" or digestive gland. Reproduction is always sexual with an individual brachiopod being either male or female, not both. Gametes are usually shed into the water where fertilization occurs. The developing larvae are free swimming for a short time and then settle to the bottom before becoming sessile as well as solitary adults.

Brachiopods are generally divided into two classes, Class Inarticulata and Class Articulata. The class Inarticulata has special muscles which hold the valves of the shell together. The shells are usually chitinophosphate making them very light and fragile. Occasionally they may be of calcium carbonate.

The class Articulata has valves with hinges of interlocking teeth, cardinal processes, and sockets as well as muscles, denoted by scars, for opening and closing the shell. They are much heavier and more robust than the Inarticulata. Their shells are always composed of calcium carbonate.

Articulate brachiopods are divided into many orders of which six are especially important: Strophomenida, Pentamerida, Spiriferida, Rhynchonellida, Orthida, and Terebratulida. Because of their abundance in the fossil record and their usefulness as index fossils, each order is briefly described.

Strophomenida is the most abundantly represented order and includes many morphologically diverse species. The shell is either flat or convex and has a wide hinge. Some of the largest and heaviest as well as most bizarre brachiopod shells are represented by this order.

Pentamerida are important index fossils in the Silurian period. They generally have a biconvex shape.

Spiriferida are noted for their unusual coiled brachidium which is frequently found intact within the fossil shell. Some species have wide hinges and others have short hinges.

Rhynchonellida have convex shells with often prominent beak, ribs, fold, and sulcus.

Orthida shells are usually biconvex, oval, and have a wide hinge and fine ribs. They were very common during the lower Paleozoic era.

Terebratulida have brachidia which are characteristically looped. The shells are generally smooth and tear shaped with a short hinge.

Chapter 9: Brachiopoda

Brachiopods were common in Paleozoic seas, especially in shallow water, and filled ecological niches which pelecypod molluscs now dominate. The Inarticulata were most diverse during Cambrian times, but the Articulata became more prominent from Ordovician to the middle Paleozoic. Abundance, as well as wide geographical range and diversity have made brachiopods very useful as index fossils.

B. Synopsis

Phylum Brachiopoda [Cambrian-Recent]

- Lophophore
- "U" shaped digestive system with both mouth and anus
- Body enclosed dorsoventrally in two unequal valves
- Marine, sessile, solitary
- Class Inarticulata [Cambrian-Recent]
  - Shell held closed by muscles only, no socket and teeth
  - Thin fragile shells, smooth, composed of chitinophosphate or calcium carbonate
- Class Articulata [Cambrian-Recent]
  - Shell held closed by muscles, have sockets, teeth, and cardinal processes
  - Shells heavier, robust, sculptured, composed of calcium carbonate
  - Order Strophomenida [Ordovician-Jurassic]
  - Order Pentamerida [Cambrian-Devonian]
  - Order Spiriferida [Ordovician-Jurassic]
  - Order Rhynchonellida [Ordovician-Recent]

Order Orthida [Cambrian-Permian]

Order Terebratulida [Devonian-Recent]

C. Representative Genera

Lingula (Inarticulata)

Ordovician-Recent

Trimerella (Inarticulata)

Silurian

Rafinesquina (Strophomenida)

Ordovician

Chonetes (Strophomenida)

Devonian-Carboniferous

Pentamerus (Pentamerida)

Silurian

Zygospira (Spiriferida)

Ordovician-Silurian

Spirifer (Spiriferida)

Mississippian-Pennsylvanian

Rhynchotrema (Rhynchonellida)

Ordovician

Rhynchotreta (Rhynchonellida)

Silurian

Terebratula (Terebratulida)

Miocene-Pliocene

Orthambonites (Orthida)

Ordovician

BRACHIOPOD MORPHOLOGY

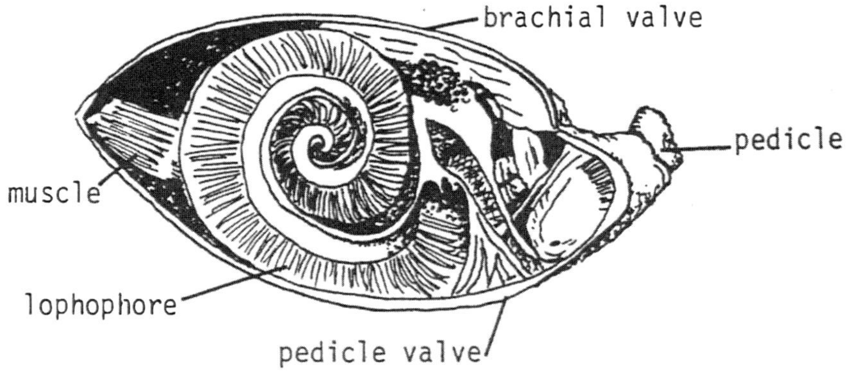

COMPARISON OF PELECYPOD AND BRACHIOPOD SYMMETRY

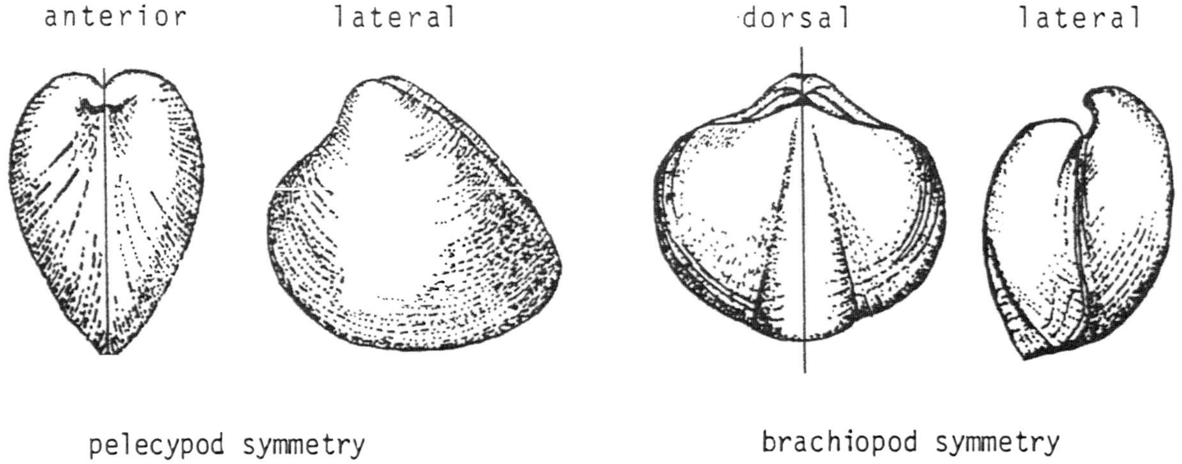

BRACHIOPOD SHELL MORPHOLOGY

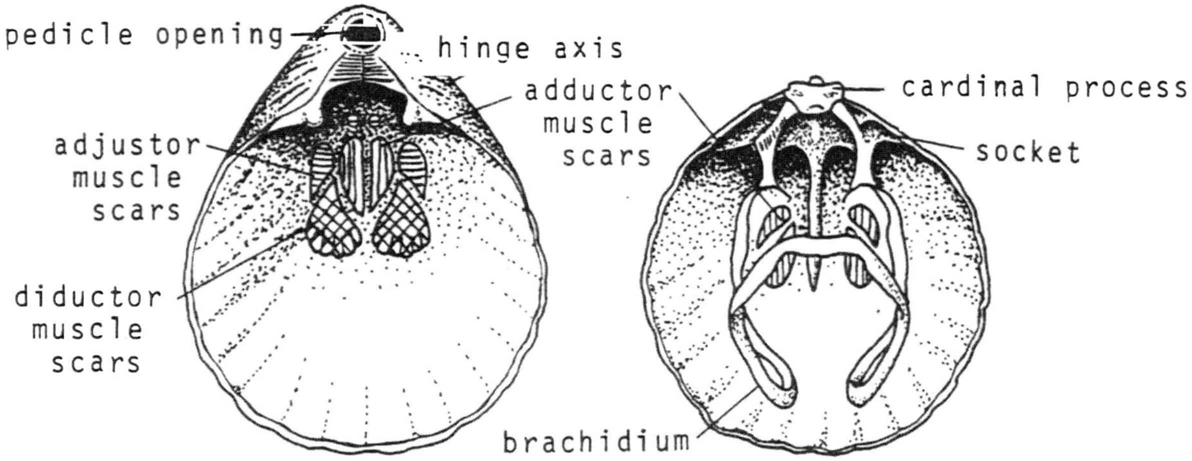

pedicle valve   brachial valve

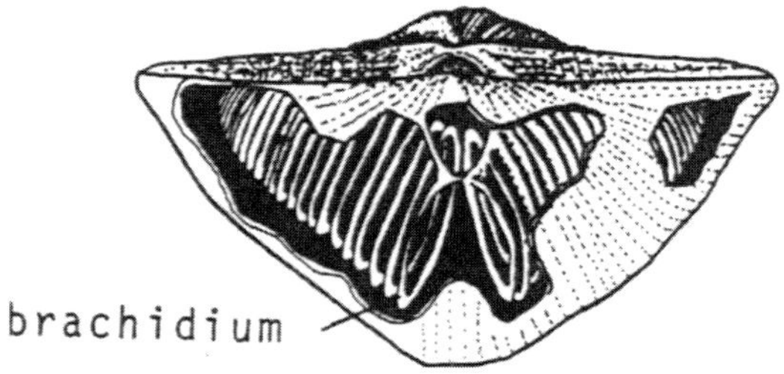

BRACHIOPODA

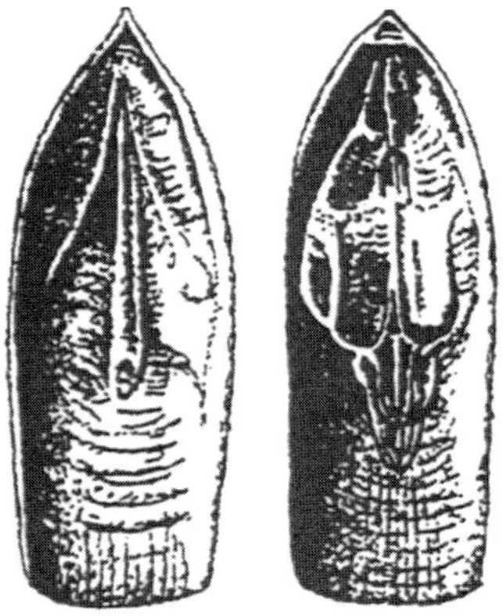

Lingula

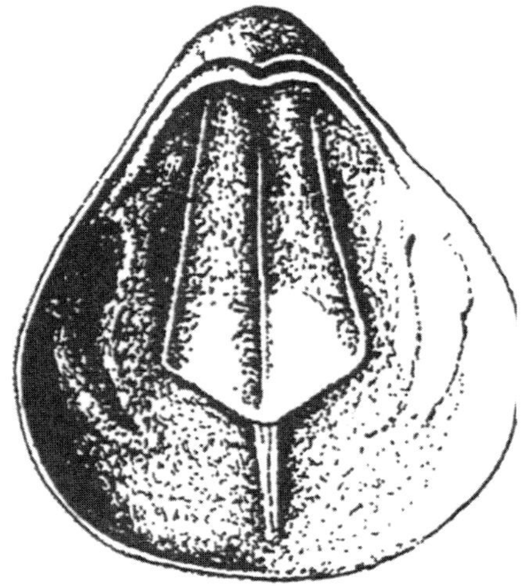

Trimerella

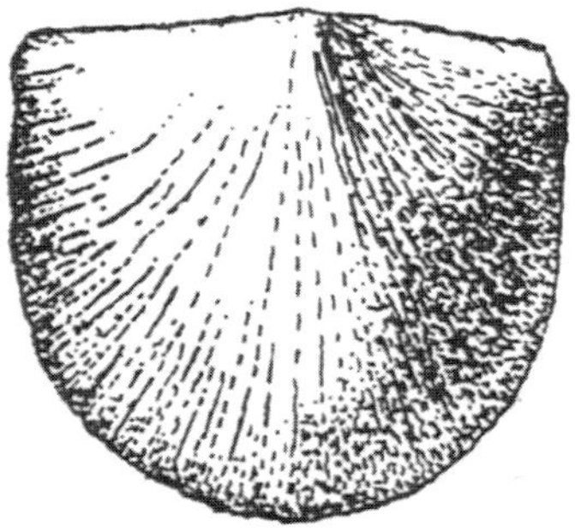

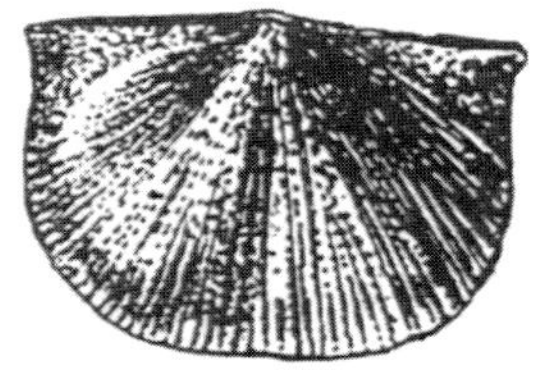

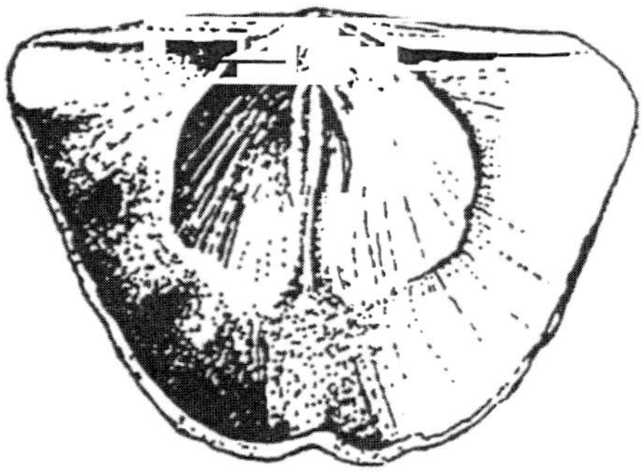

Rafinesquina

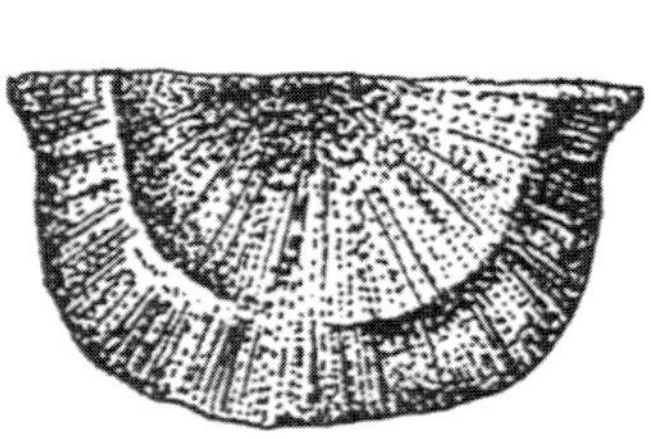

Chonetes

Pentamerus

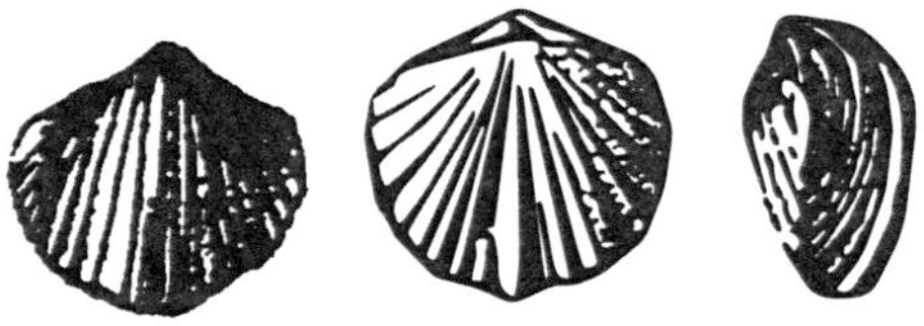

Zygospira

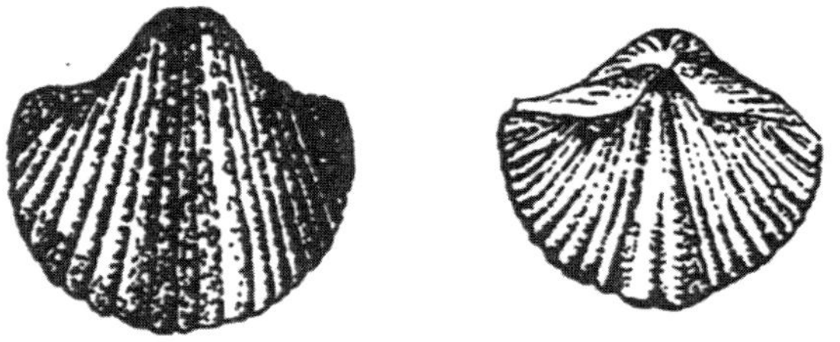

Spirifer

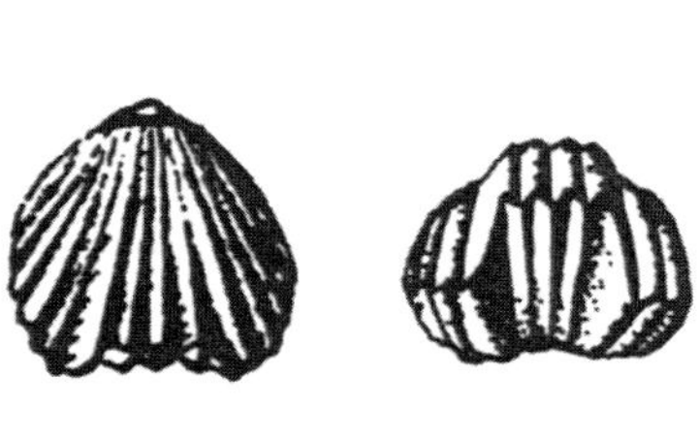

Rhynchotrema

Rhynchotreta

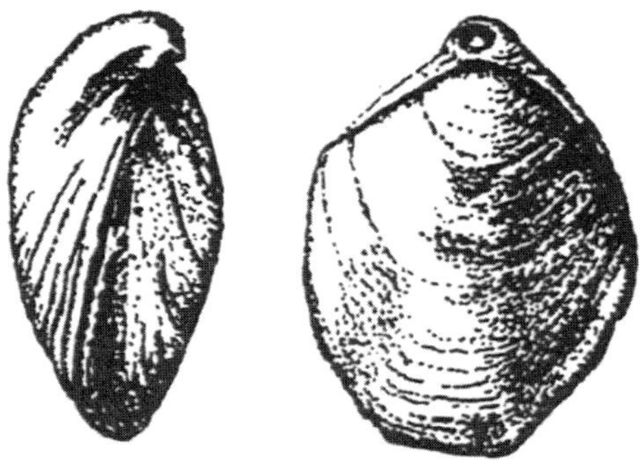

Terebratula

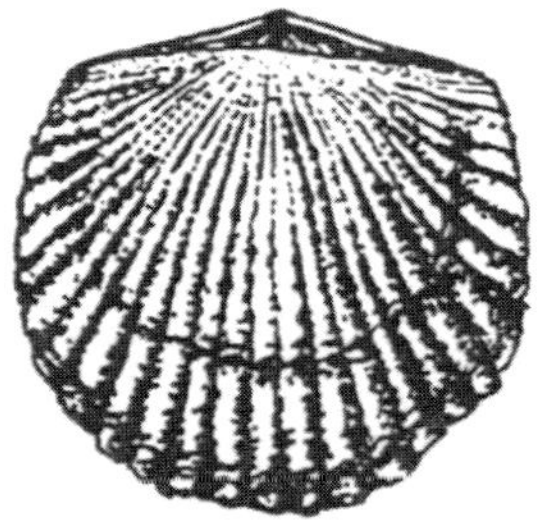

Orthambonites

NOTES

Chapter 10: MOLLUSCA

A. Discussion

The phylum Mollusca contains an abundance of species including various types of snails, clams, chitons, limpets, squid, and abalone. Although differing life-styles have greatly modified the body plan of each class, basic features remain.

Molluscs are bilaterally symmetrical organisms with well developed tissues and organs. They have digestive systems which include both a mouth and an anus, an excretory system, a primitive heart, paired gills for respiration (some species have secondarily lost these), and a well developed nervous system. Most species also have a radula or rasping organ located in their mouth, which is used to feed, scrape food off a surface, or drill through the shell of prey. The majority of species have calcareous shells, secreted by a mantle, which provide structure, support for muscles and organs, and protection. All species also have a muscular foot for locomotion, although in cephalopods it has been modified into tentacles.

There are six classes of molluscs which we will review: Monoplacophora, Polyplacophora, Scaphopoda, Gastropoda, Pelecypoda, and Cephalopoda.

Monoplacophora are thought to represent one of the more primitive body arrangements of the Mollusca. A cap shaped shell covers the soft segmented body and mantle. The mouth is anterior and the anus is posterior. This seemingly obvious arrangement becomes more significant when it is observed that in more developed classes, rotation has moved the anus to an anterior position. Monoplacophorans for many years were thought to be extinct, however, living specimens found in the middle of this century have revised scientific thinking. One curious aspect of these recent specimens has been the discovery of multiple pairs of nephridia, muscles, and gills, characteristics not shown by other classes. Such examples of metamerism or segmentation have increased interest and speculation as to the possible ancestors of molluscs.

Polyplacophorans or amphineurans have a shell which is usually divided into eight overlapping plates covering a flattened oval body. Like monoplacophorans, the mouth is anterior and the anus is posterior. They have a broad flat foot and much reduced head with accompanying reduction in sensory organs. These modifications are thought to have accompanied the polyplacophoran life style which is sluggish and herbivorous. Chitons, familiar inhabitants of tidepools, are polyplacophorans. They are generally found as disarticulated plates in deposits from Cambrian to recent times. The fossils are relatively rare and, consequently, not especially useful as biostratigraphic indicators.

Scaphopoda are marine molluscs which have an elongated slightly curved shell, somewhat like a tusk, which is open at both ends. Although they do not show great modifications in their body plan from previous molluscs discussed, scaphopods have secondarily lost their gills. They are burrowing marine organisms, living head downward in the sand with the smaller posterior part of the shell extended above the substrate. They have a long fossil record, remaining unchanged from the Ordovician period to the present. Their lack of diversity as well as abundance prevent them from being very useful as index fossils.

Gastropods make up the largest class of molluscs, in part due to their extensive radiation. They have adapted to many environments from fresh to marine water and even have members which are terrestrial. In conjunction with their ecological radiation, the food habits of gastropods are very diversified and range from herbivorous to carnivorous.

The most significant change in the gastropod body plan from the molluscs discussed earlier is the torsion or rotation of the body so that the anus lies in an anterior position somewhat above the mouth. The body is frequently enclosed in a spiral shell. The gastropod foot is prominent as is the head with its accompanying sensory organs. Although aquatic species have gills, the terrestrial species have lost theirs with an accompanying modification of the mantle into a lung for respiration. Gastropods make up an abundant and very significant part of the fossil record from the Cambrian period to the present. Their abundance, range, and diversity could make them very useful biostratigraphic indicators as well as paleoenvironmental indicators; however, a scarcity of workers on fossil gastropods has hindered their usefulness.

Pelecypods are clams, mussels, scallops, and oysters. They have been given many names in the past including Lamellibranchia and Bivalvia as well as Pelecypoda. Their shells consist of two identical dorsally hinged valves which enclose the body laterally. Remember that brachiopods also have two valves, but these enclose the body dorsoventrally and are not identical. As noted in an illustration in the brachiopod chapter, a line of symmetry would go through both valves in Brachiopoda but would go between the valves in Pelecypoda.

The usual molluscan radula used in feeding has been secondarily lost in the pelecypods and gills have become modified for food gathering as well as for respiration. As water is passed over the gills, particles of food are filtered out and oxygen is removed. Other modifications have taken place in the pelecypods which allow them to burrow deep into the substrate. These modifications include a digging foot which is laterally compressed and an enlarged mantle attached to the shell along the pallial line. The attachment of the mantle and shell prevents sand and debris from getting into the interior of the shell and damaging the soft body of the clam. In a few species the mantle has undergone additional modifications, becoming fused as inhalent and exhalent siphons facilitating the movement of water in and out of a clam which is deeply buried. Most pelecypods are not attached to the substrate although several exceptional species have developed special mechanisms for attachment to hard substrates. Mussels attach by means of unique byssus threads and oysters have evolved a technique for cementing themselves to suitable surfaces.

Fossil pelecypods occur abundantly throughout the world from the Cambrian period to the present but are not generally an important tool for correlation or biostratigraphy until Cenozoic time. They do, however, provide an indication of ancient environments reflecting whether the substrate was sand, mud, or rock.

Class Cephalopoda includes many fossil species as well as the modern octopus, squid, and cuttlefish. They are the largest and most advanced of the invertebrates with an especially well developed nervous system. Their morphological development reflects a predatory life-style requiring great mobility and well developed sensory organs. For the most part cephalopods have relinquished the bottom dwelling habits of other molluscs and pursue their prey in a pelagic environment.

The shell of a cephalopod, when present, has transverse septa dividing it into internal chambers. As the organism grows, it secretes new shell material around the edges of the opening and seals off the shell behind it with a septum. The septa are joined by a siphuncle or tube which allows gas and liquids to enter and leave the chambers regulating shell buoyancy. The shells may be straight or coiled. Cephalopod bodies differ from other molluscs by having the foot modified into tentacles which serve both for locomotion and for catching prey.

There are several groups of cephalopods including Nautiloidea, Ammonoidea, and Coleoidea. The Nautiloidea evolved during the Cambrian period and were important throughout the Paleozoic era. Their septa are flat or gently curved. The septa of Ammonoidea are complex and make intricate patterns where they attach to the inside of the shell. The patterns or sutures are the basis for distinguishing different species of ammonoids. They have names designating their complexity with goniatite being one of the simpler patterns and ammonite one of the more complex. Ammonoids developed in the Devonian period and were very important during the Mesozoic era before becoming extinct in the Cretaceous. Their short biostratigraphic ranges make them very useful index fossils.

Order Coleoidea contains most of the living species of cephalopods. Their shells are either absent or exist as much reduced internal structures. As fossils, usually only these internal shells are found. These can be bullet shaped structures, called phragmocones, which are important enough in some cases to serve as index fossils.

B. Synopsis

Phylum Mollusca [Cambrian-Recent]

True tissues and organs, digestive system with both mouth and anus

Radula and calcareous shells in most species

Class Monoplacophora [Cambrian-Recent]

Cap like shell

Class Polyplacophora (Amphineura) [Cambrian-Recent]

Shells of eight articulating plates, broad foot, sense organs reduced

Class Scaphopoda [Devonian-Recent]

Slightly curved tusk like shell, open at both ends, no gills

Class Gastropoda [Cambrian-Recent]

Shell may be spirally coiled, anus in anterior position

Chapter 10: Mollusca

Class Pelecypoda [Cambrian-Recent]

Two valves enclose body laterally, no radula, modified gills, siphons in some species

Class Cephalopoda [Cambrian-Recent]

Foot modified into tentacles, highly developed eyes and sensory organs

Chambered shell

Order Nautiloidea [Cambrian-Recent]

Shells may be straight or coiled, septa flat or gently curved

Order Ammonoidea [Ordovician-Cretaceous]

Septa complex: goniatite, ceratite, ammonite

Order Coleoidea [Carboniferous-Recent]

Shells absent or much reduced internal structures

Suborder Belemnoidea [Carboniferous-Cretaceous]

Chambered, conical internal shell called phragmocone

Chapter 10: Mollusca

C. Representative Genera

Chiton (Polyplacophora)

Cambrian-Recent

Dentalium (Scaphopoda)

Triassic-Recent

Bellerophon (Gastropoda)

Silurian-Triassic

Turritella (Gastropoda)

Oligocene-Recent

Worthenia (Gastropoda)

Mississippian-Triassic

Haliotis (Gastropoda)

Miocene-Recent

Acmaea (Gastropoda)

Oligocene-Recent

Ostrea (Pelecypoda)

Triassic-Recent

Mytilus (Pelecypoda)

Triassic-Recent

Pecten (Pelecypoda)

Eocene-Recent

Unio (Pelecypoda)

Triassic-Recent

Nautilus (Cephalopoda)

Oligocene-Recent

Chapter 10: Mollusca

Orthoceras (Cephalopoda)

Ordovician

Goniatites (Cephalopoda)

Mississippian

Ceratites (Cephalopoda)

Triassic

Phylloceras (Cephalopoda)

Jurassic-Cretaceous

Baculites (Cephalopoda)

Cretaceous

Belemnoidea

Jurassic-Eocene

MOLLUSCA MORPHOLOGY

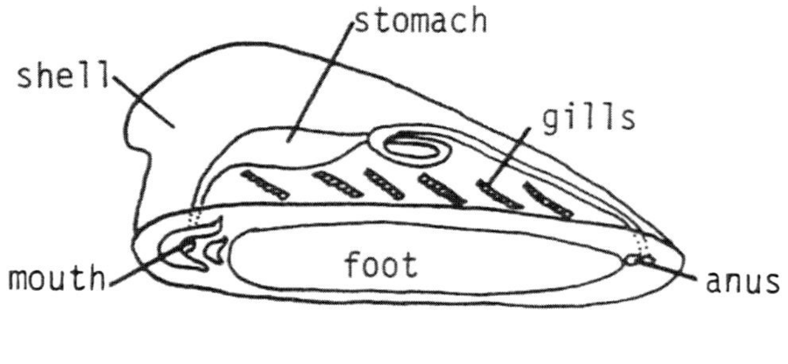

MONOPLACOPHORA

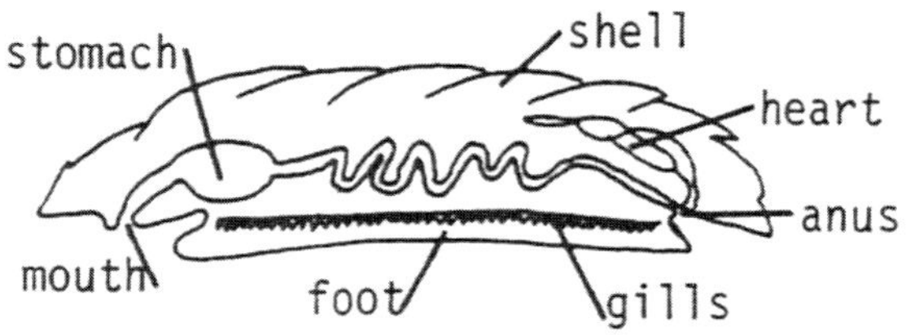

POLYPLACOPHORA

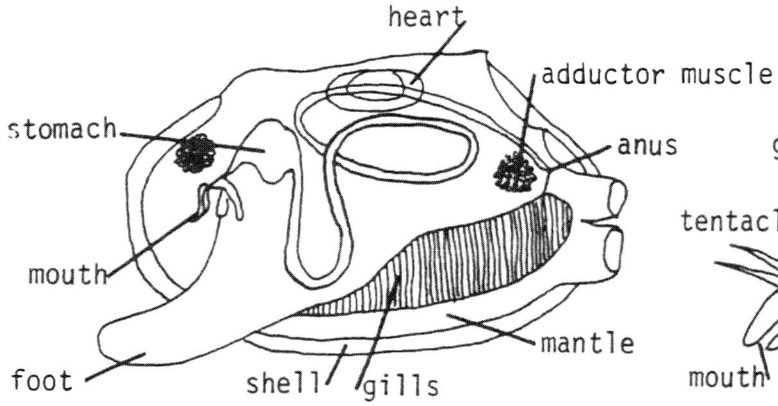

PELECYPODA

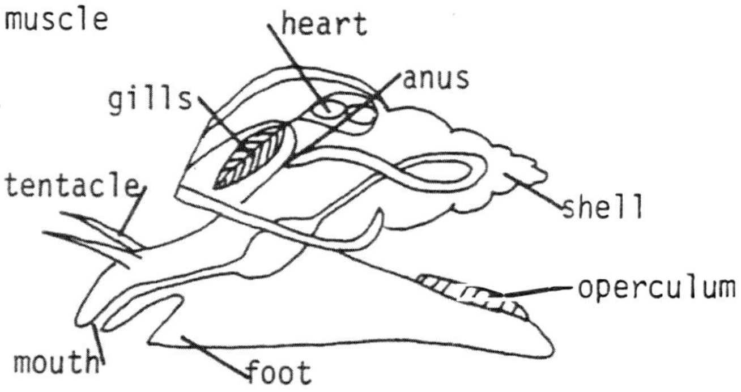

GASTROPODA

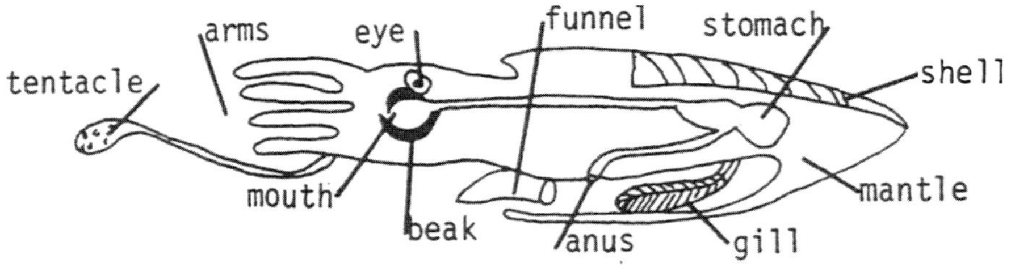

CEPHALOPODA

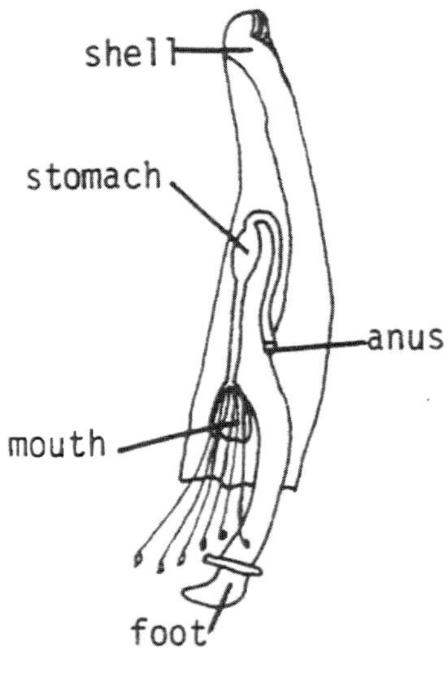

SCAPHOPODA

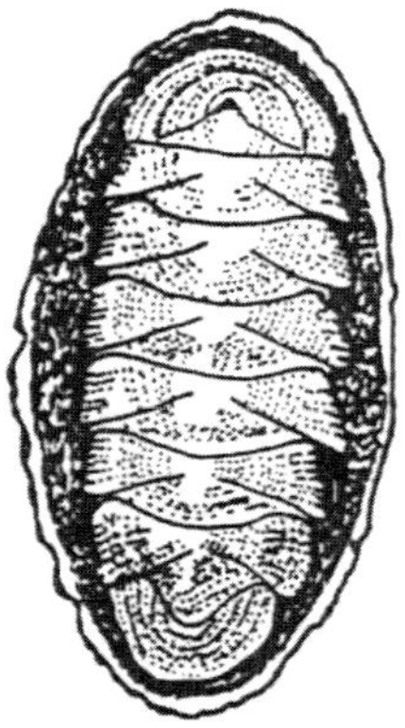

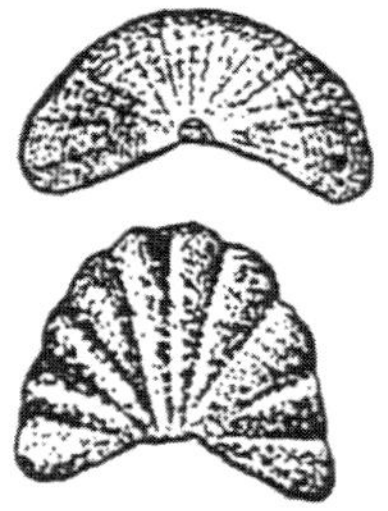

chiton

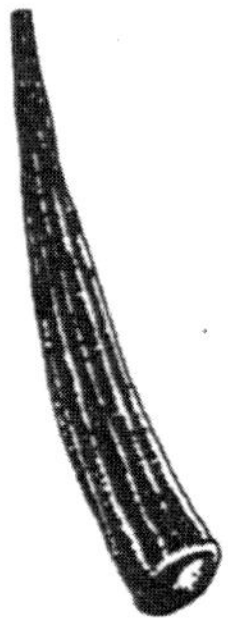

Dentalium

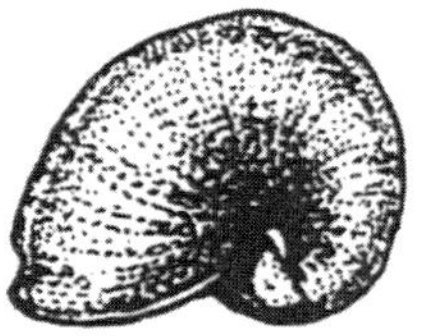

Bellerophon

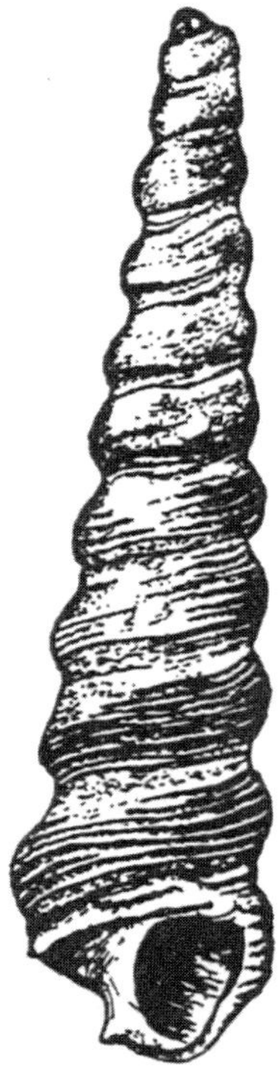

Turritella

Worthenia

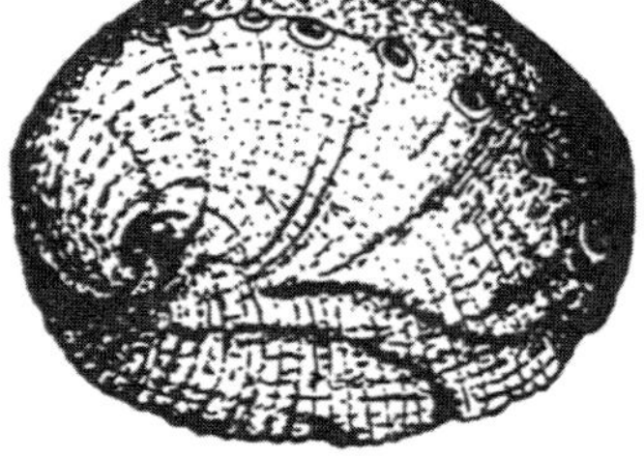

Haliotis

Acmaea

PELECYPODA

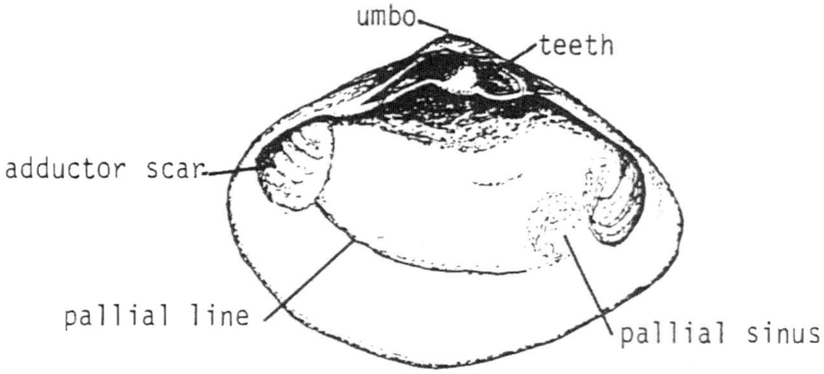

interior of shell

Ostrea

SHELL MORPHOLOGY

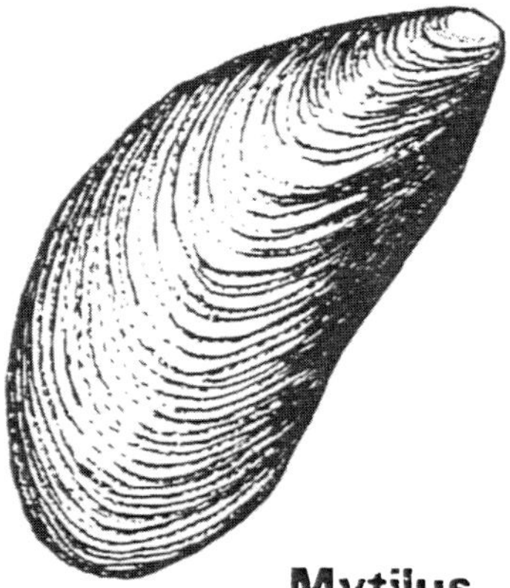

Mytilus

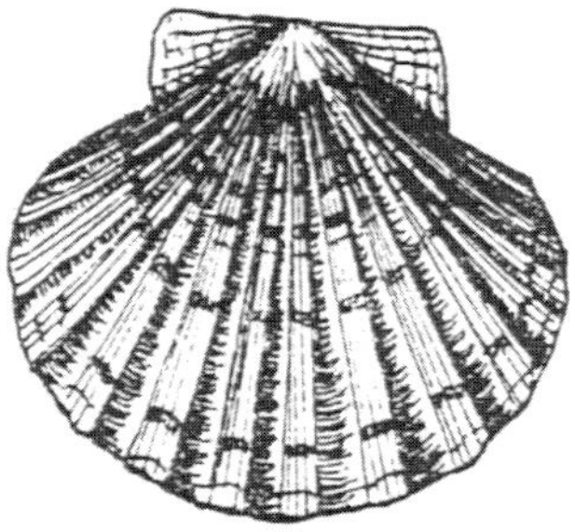

Pecten

Unio

CEPHALOPODA

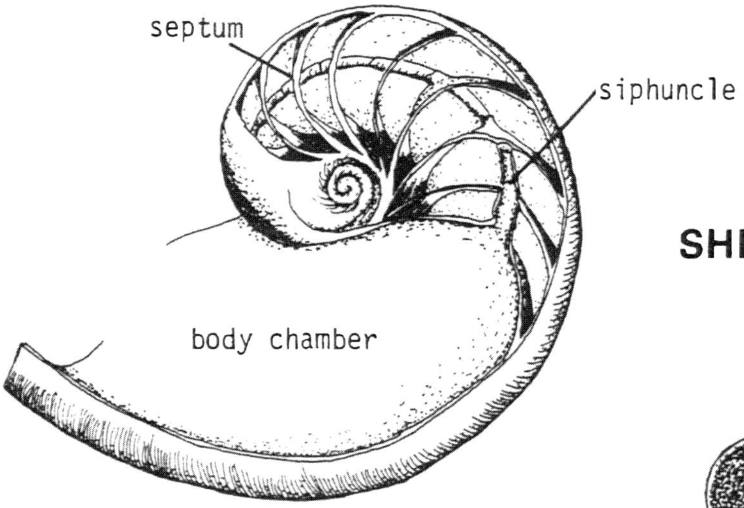

SHELL MORPHOLOGY

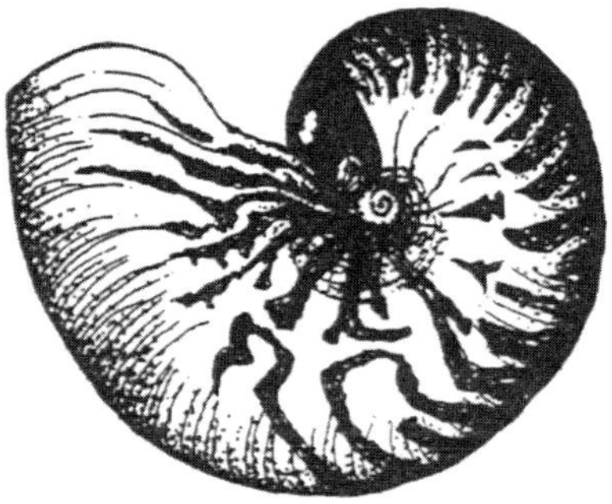

Nautilus

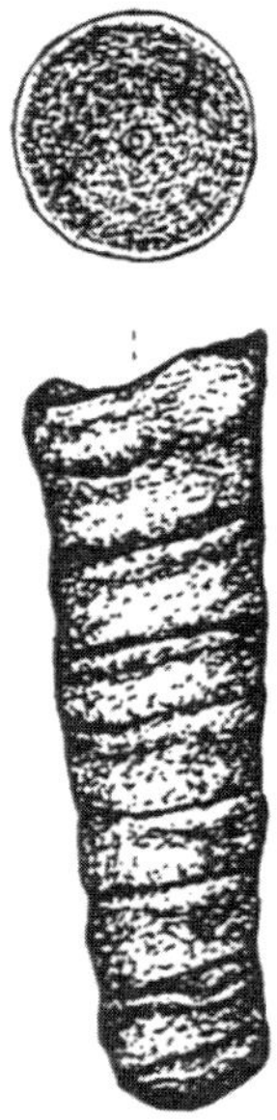

Orthoceras

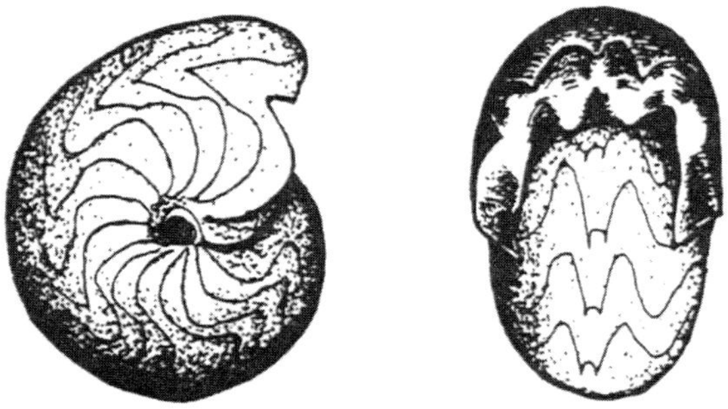

Goniatites

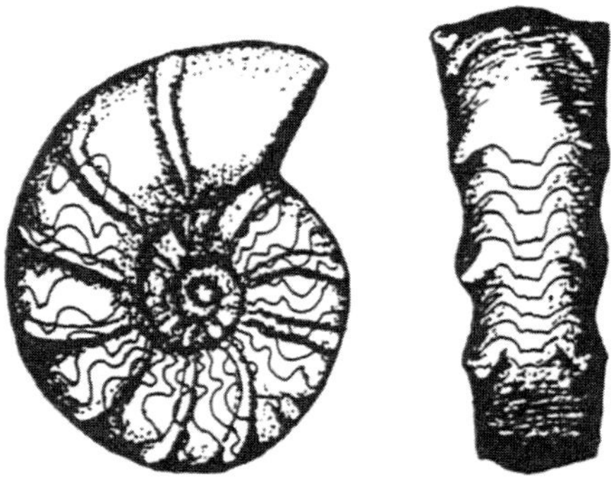

Ceratites

CEPHALOPODA

Baculites

surface of septum

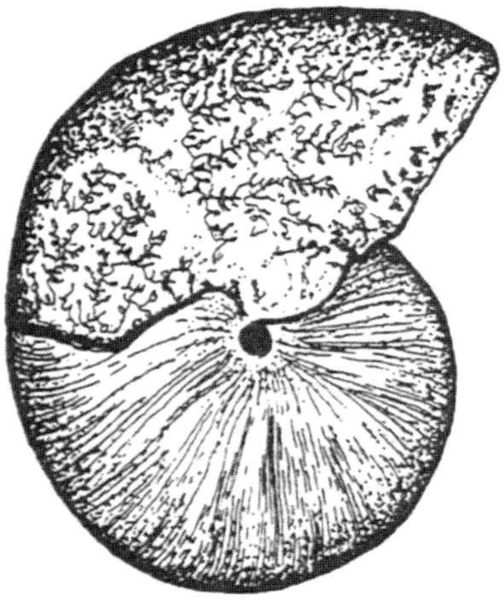

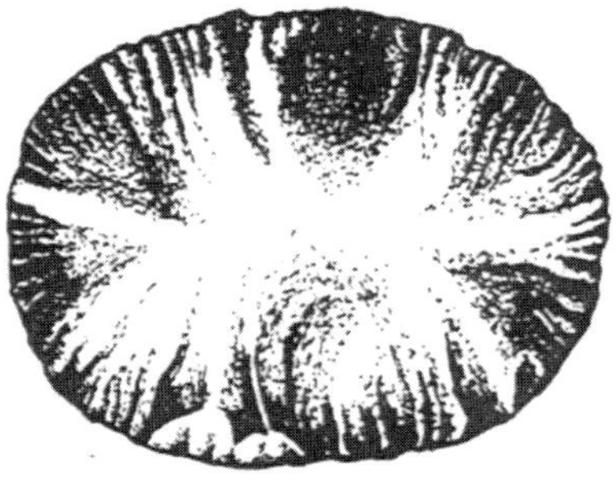

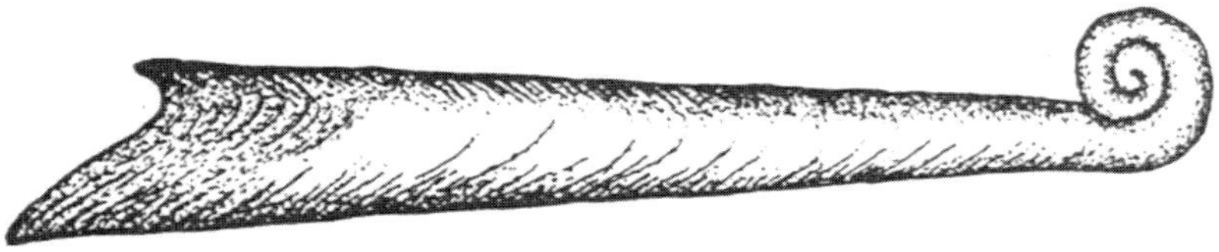

ammonite suture pattern

Phylloceras

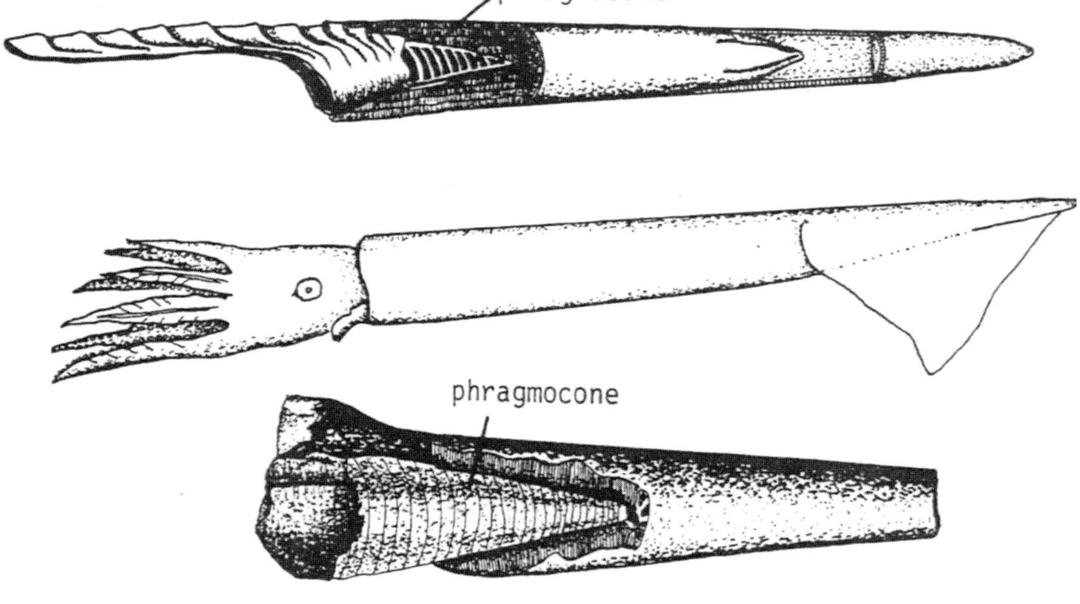

Belemnoidea

NOTES

Chapter 11: ARTHROPODS

A. Discussion

Arthropods are the most abundant group of animals in the world, having successfully adapted to numerous ecological niches on land, in water, and in the air. Current opinion is divided between a single phylum and a multi-phylum approach to the classification of arthropods. Instead of calling arthropods a phylum we will use the term by itself without assigning a special role in the classification scheme. We mean the term to include among others insects, crustaceans, arachnids, trilobites, and eurypterids.

The arthropod body, unlike that of previously discussed phyla, is segmented, much like the annelid worms. The shape is bilaterally symmetrical and generally elongated with paired jointed appendages. Surrounding the body is a chitinous exoskeleton which is periodically shed or molted as the animal grows. The head region and sensory organs are well developed, and reflect a mobile life-style. Arthropods have an open ciculatory system which includes a heart. They also have advanced excretory, nervous, and reproductive systems. The sexes are typically dioecious, separate males and females, with often elaborate mechanisms for fertilization. One such device is a specially modified appendage which many crustaceans possess for transferring sperm to a receptive female.

The most familiar group of arthropods to us are the insects, the most abundant group of animals on earth. They will not be covered here because they are, with few exceptions, poorly and infrequently preserved and therefore not commonly encountered in the fossil record. The groups of arthropods of most interest to paleontologists are Trilobita, Merostomata, and Crustacea.

Class Trilobita contains the most distinctive arthropods. They thrived in Cambrian and Ordovician seas, becoming one of the dominant groups in the fossil record. Nevertheless, they became extinct by the end of the Permian period. The body is typically a flattened oval divided into frontal (cephalon), middle (thorax), and posterior (pygidium) sections. Trilobites are divided lengthwise into three lobes, which give them their name. The lobes consist of a central axial lobe and two lateral pleural lobes. Numerous biramous appendages extend laterally from the trilobite body. The thorax or trunk has two or more freely moveable segments. The segments in some trilobites are so flexible that the fossils are occasionally found rolled up like a sow bug with the pygidium tucked under the cephalon. Trilobite exoskeletons generally had minerals impregnating the citin, thus they were readily preserved. Large species populations and the potential of several molted exoskeletons per individual have left an abundance and diversity of trilobites, making them excellent index fossils for the Paleozoic era.

Class Merostomata includes the horseshoe crabs as well as the extinct sea scorpion or Eurypterid. They lived in marine, fresh, and brackish waters throughout the world, particularly in the Paleozoic. The body is composed of a prosoma or head region, and opisthosoma or trunk region. Compound eyes and median ocelli are present on the dorsal surface. Six pairs of jointed appendages extend laterally from the prosoma. Some of the appendages are walking legs, some are modified for swimming, and some evolved into grasping appendages. Associated with the mouth on the underside of the prosoma, the first pair of appendages are piercing and tearing organs called chelicerae. Eurypterids, which were as long as nine feet, had both grasping appendages and formidable chelicerae which probably made them fierce predators. Merostomata are not particularly abundant in the fossil record, but occasionally make up a large part of certain fossil assemblages. Some species are useful indicators of quiet estuarine or brackish environments.

Phylum Crustacea includes barnacles, crabs, lobsters, shrimp, and crayfish. They are found in all waters from fresh to marine and also include some terrestrial forms. Crustaceans differ from other classes of arthropods in having two pairs of antennae, a pair of mandibles, and two pairs of maxillae. Sometimes a shield developes posterior of the head, termed a carapace. Depending on the group, appendages are variously developed. The trunk region is frequently divisible into an anterior thorax and posterior abdomen. Only two classes of Crustacea are considered here: Ostracoda and Cirripedia, which have particular fossil significance.

Ostracoda have evolved a carapace which is greatly enlarged and extends over the entire body as a bivalved shell. It is impregnated with calcium carbonate, which provided rigidity and has also enabled them to be well represented in the fossil record. Growth occurs by molting, as in the other arthropods, and consequently immature molt stages are common as fossils. Ostracods occur in marine, fresh, or brackish water with the heavier, more sculptured shells generally indicating a marine environment, and fragile thinner shells reflecting brackish or fresh water. Paleontologists have found ostracods useful for stratigraphic correlation because of their abundance, diversity, and short ranges. They frequently occur with Foraminifera and together are important for biostratigraphic dating.

Class Cirripedia contains the barnacles, which are the only sessile crustaceans. In adapting to a sessile existence, barnacles have lost many gross morphological crustacean features, and were initially thought to be molluscs. A careful look at the internal anatomy and larval stages clarified their position in the animal kingdom. Two types of barnacles are recognized, stalked, and unstalked types. The stalked variety are frequently referred to as goose necked barnacles, the unstalked as acorn barnacles. In both types of barnacles, calcareous plates secreted by a mantle surround the living animal. An exoskeleton, which is molted during growth, lines the interior of the mantle cavity and encloses the appendages. Two plates, the tergum and scutum, act as doors, which open to allow the barnacles' appendages, or cirri, to entrap food. The tergum and scutum can be closed for protection or to prevent dessication during low tides. Other plates, (rostrum, carina, and various kinds of laterals) protect the main part of the body. These plates form the capitulum, and sit on top of a stalk or peduncle which is usually present. Cirripedia are found in the fossil record most often as disarticulted calcareous plates. They generally are not very abundant, but are proving to be increasingly useful for correlation and biostratigraphy especially in the Cenozoic.

B. Synopsis

Arthropods [Cambrian-Recent]

Segmented, with paired jointed appendages, ocelli or compound eyes

Exoskeleton of chitin which may be impregnated with calcium carbonate

Growth by molting

Phylum Trilobitomorpha [Cambrian-Triassic]

Class Trilobita [Cambrian-Permian]

Axial and pleural lobes

Cephalon, thorax, and pygidium

Phylum Cheliceriformes [Cambrian-Recent]

Class Merostomata [Cambrian-Recent]

Six pairs of appendages modified for tearing, grasping, walking, swimming

Body divided into two regions, prosoma, opisthomsoma

Phylum Crustacea [Cambrian-Recent]

Frequently with posteriorly directed head shield or carapace

Two pairs of antennae, mandibles, two pairs of maxillae

Class Ostracoda [Cambrian-Recent]

Carapace expanded to form hinged calcareous bivalved shell

Class Cirripedia [Cambrian-Recent]

Calcium carbonate plates

Chapter 11: Arthropods

C. Representative Genera

Agnostus (Trilobita)

Cambrian

Olenellus (Trilobita)

Cambrian

Phacops (Trilobita)

Silurian-Devonian

Dionide (Trilobita)

Ordovician

Mixopterus (Merostomata)

Devonian

Eurypterus (Merostomata)

Ordovician-Carboniferous

Limulus (Merostomata)

Recent

Savagellites (Ostracoda)

Mississippian

Knightina (Ostracoda)

Mississippian-Permian

Quadrijugator (Ostracoda)

Ordovician

Eurychilina (Ostracoda)

Ordovician-Silurian

Balanus (Cirripedia)

Eocene-Recent

Gooseneck Barnacle

Silurian-Recent

ARTHROPOD MORPHOLOGY

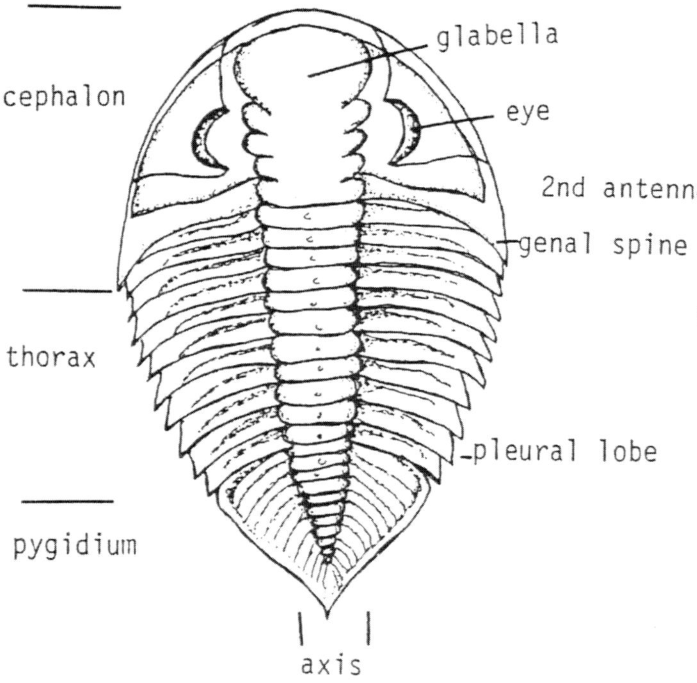

TRILOBITE

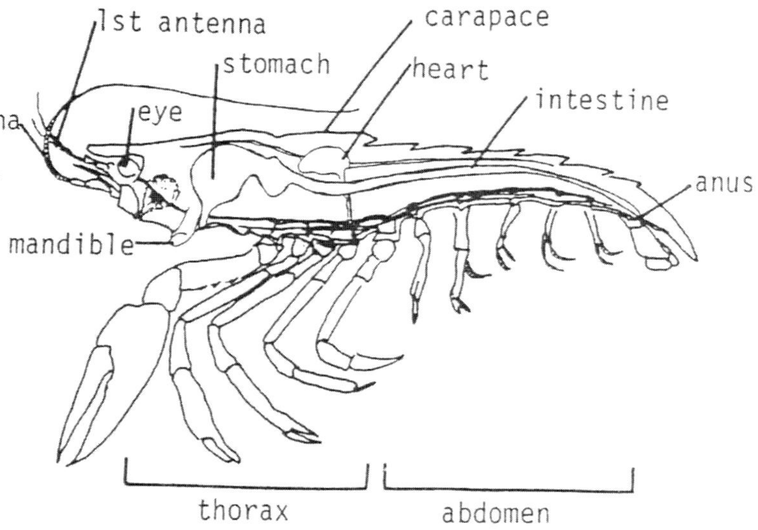

CRAYFISH

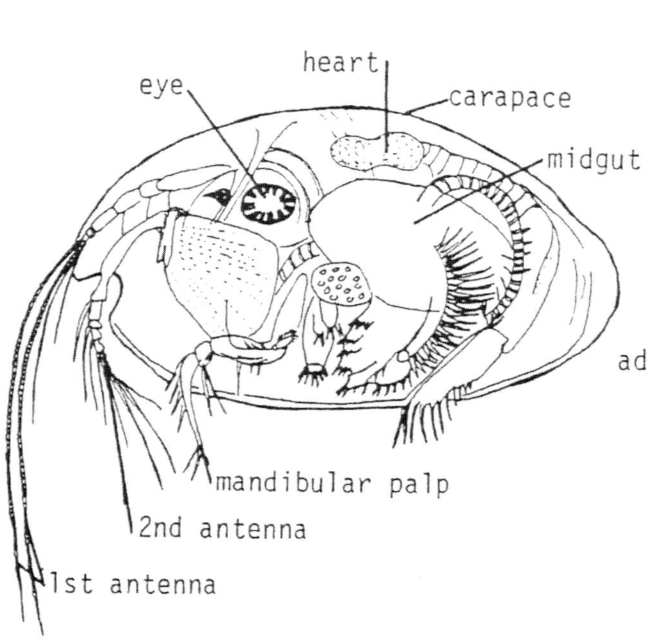

OSTRACOD

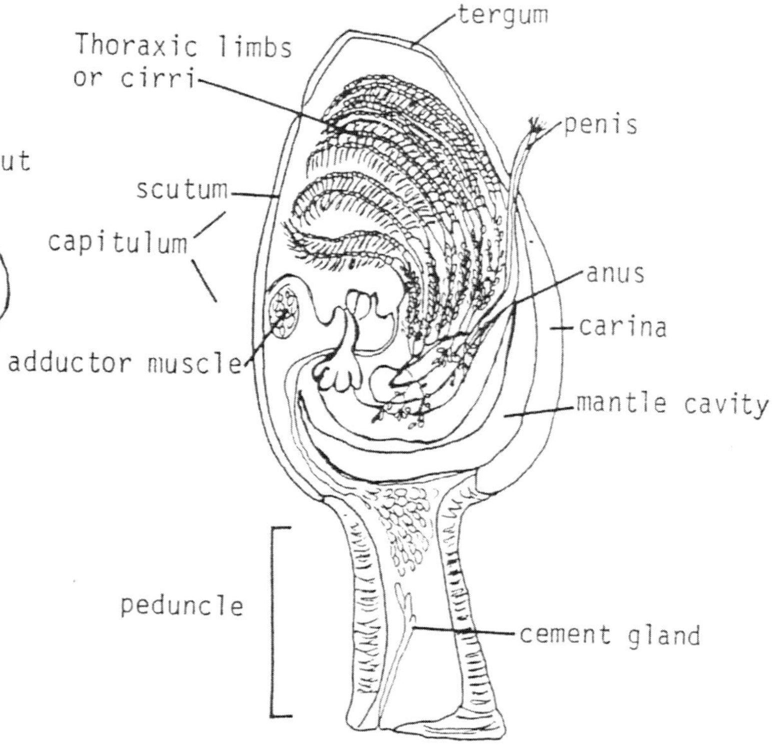

CIRRIPEDIA

(After Treatise)

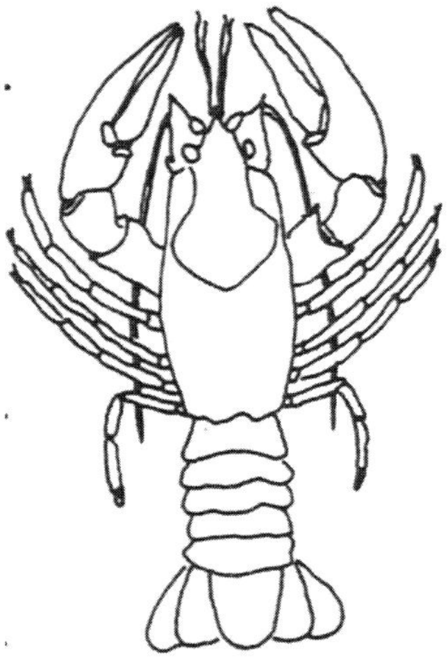

CRUSTACEA

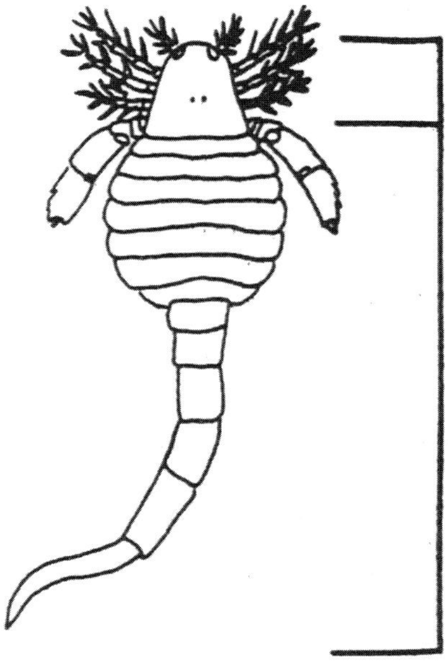

MEROSTOMATA

(After Treatise)

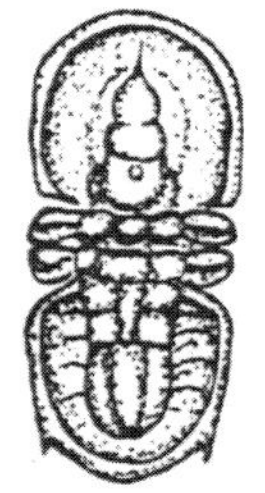

Agnostus

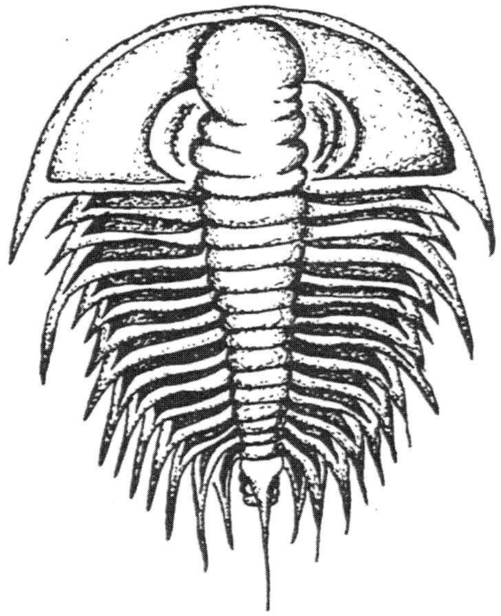

Olenellus

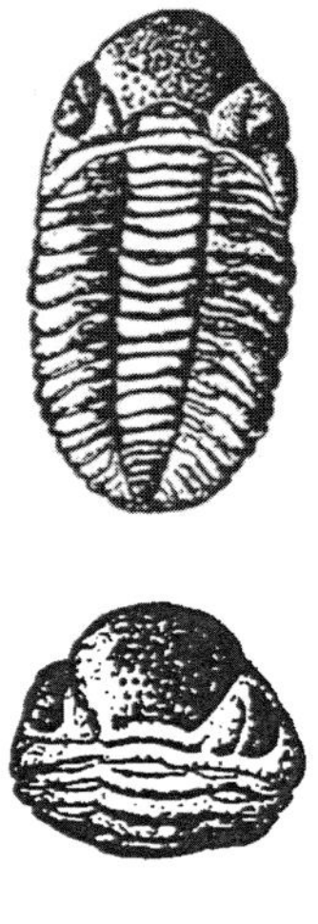

Phacops

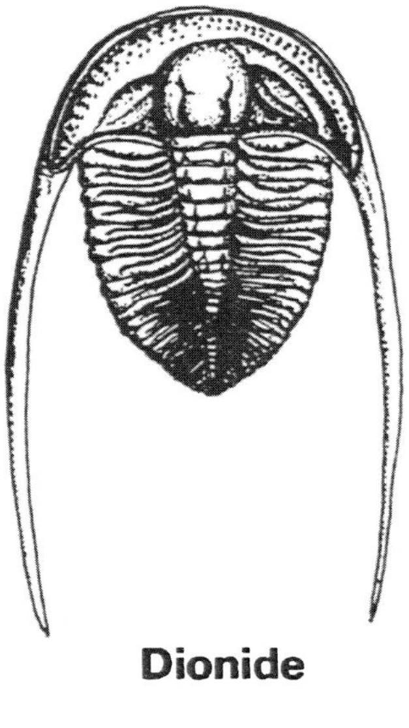

Dionide

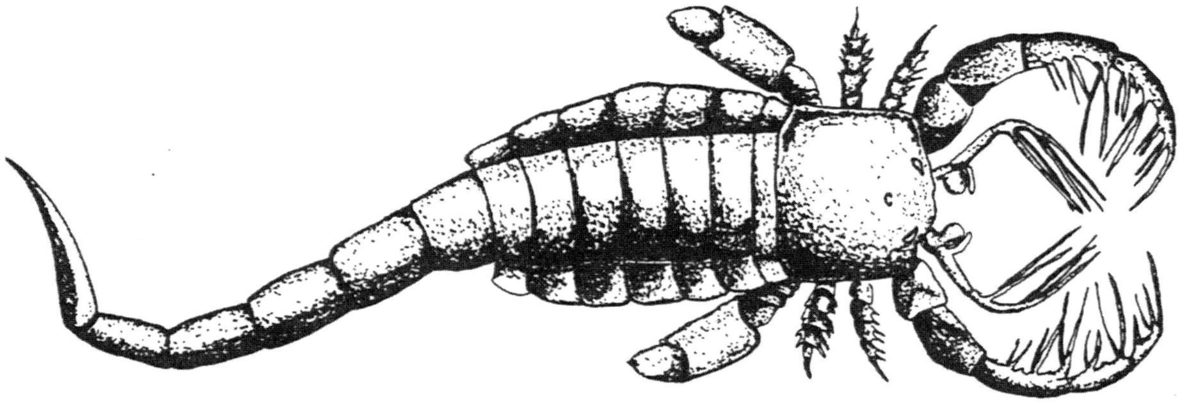

Mixopterus

(After Treatise)

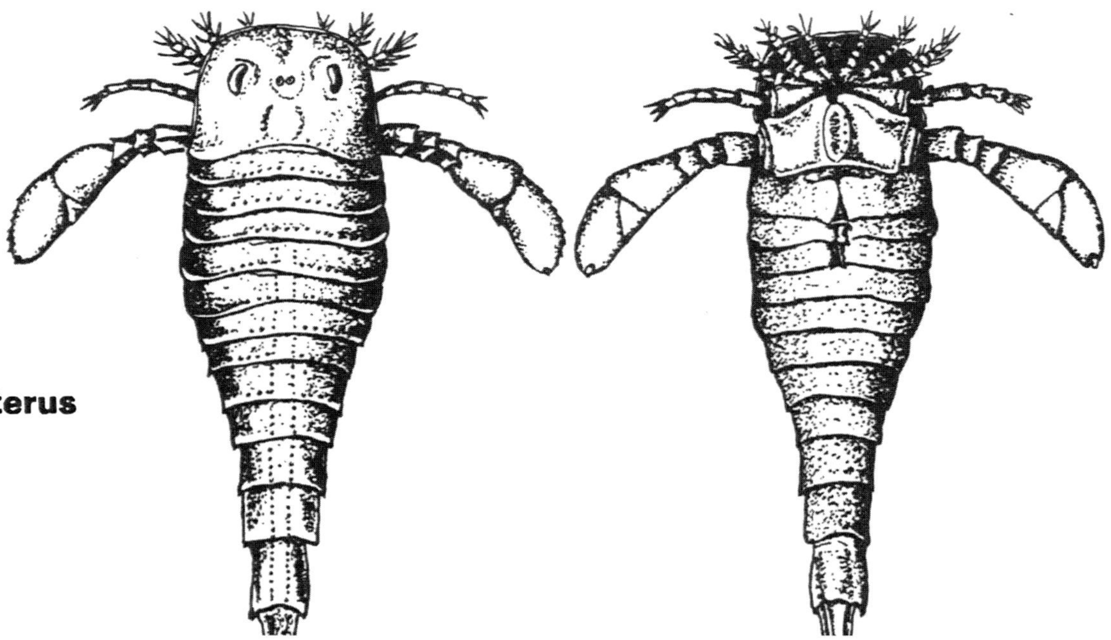

Eurypterus

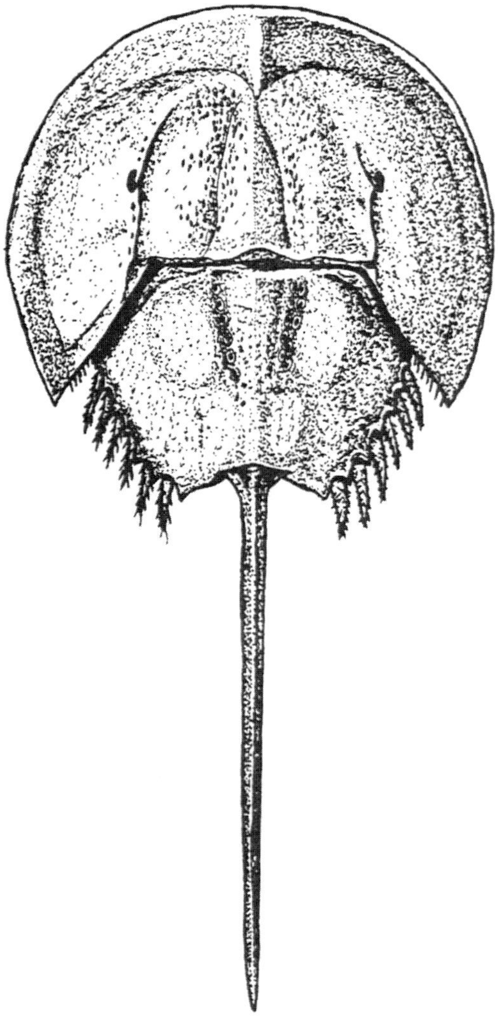

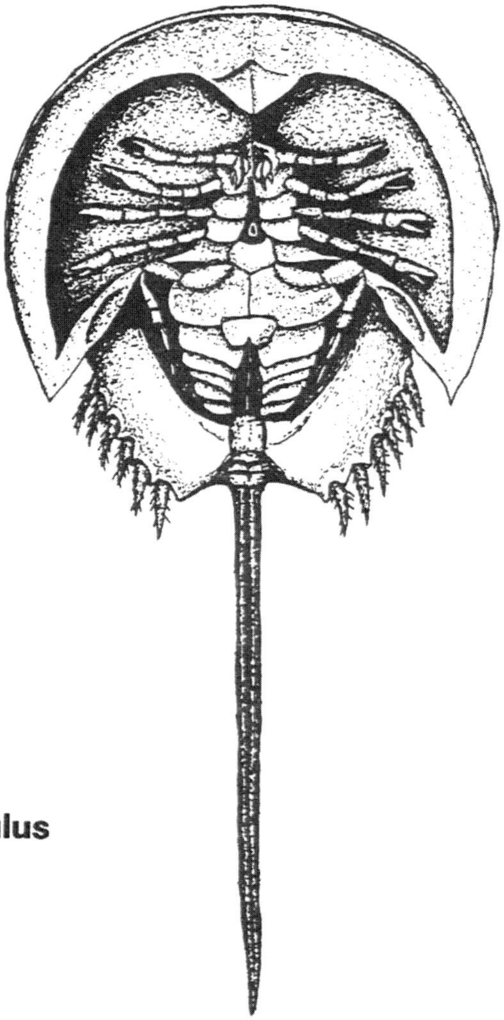

Limulus

OSTRACODA

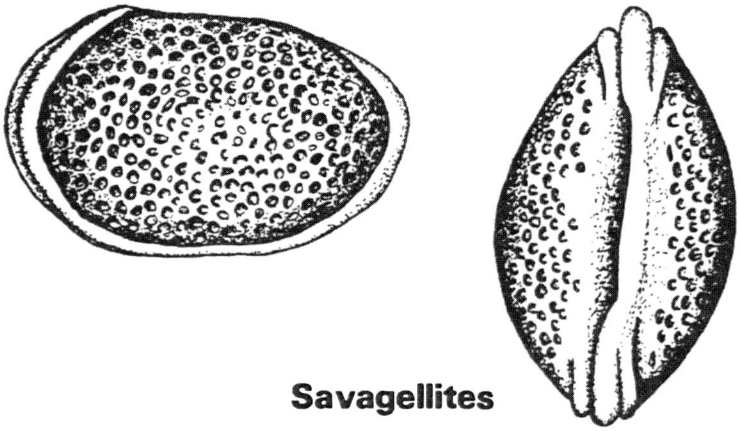

Savagellites
(After Treatise)

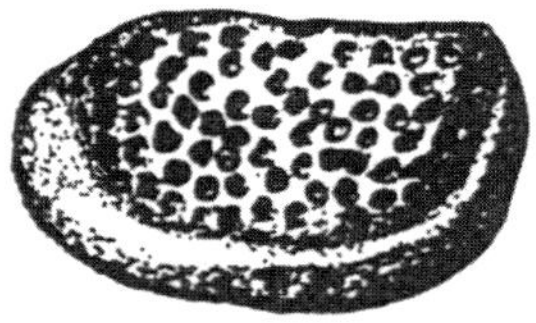

Knightina

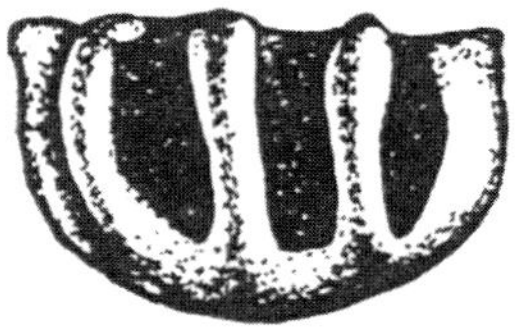

Quadrijugator

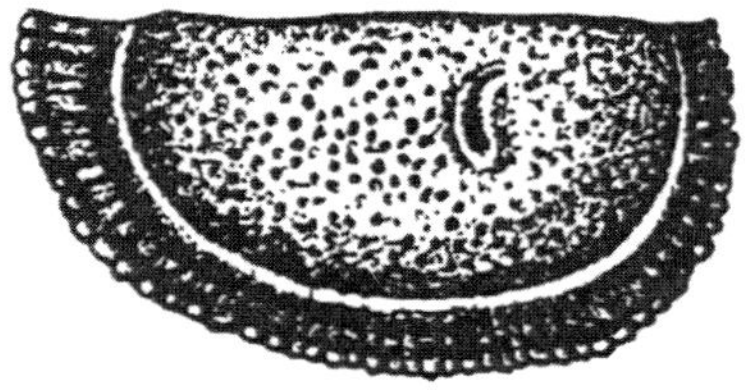

Eurychilina

Balanus

gooseneck barnacle

NOTES

Chapter 12: ECHINODERMATA

A. Discussion

Phylum Echinodermata are the "spiny skin" animals such as sea urchins, sand dollars, sea lilies, and sea cucumbers which live on the ocean floors throughout the world.

Most echinoderm adults are characterized by radial symmetry while the larval stages are bilaterally symmetrical. Compared to other phyla showing radial symmetry, they are very advanced, having a true coelom and well developed organ systems. One organ complex, unique to the phylum, is called the water vascular system, a special arrangement of rings and canals containing fluid which is pumped from one part of the system to another. Some of the canals extend into tube feet which lie along the edges of channels called ambulacral grooves. The movement of fluid through the water vascular system acts as a hydraulic pump which moves the tube feet, allowing them to function both in locomotion and feeding.

Chapter 12: Echinodermata

Echinoderms have a well developed reproductive system. They are dioceious, that is, the sexes are separate; and gametes are released directly into the water where fertilizaion occurs. The endoskeleton of echinoderms consists of plates of calcium carbonate having the same physical structure as crystalline calcite. These plates are a diagnostic characteristic of the phylum. Often the plates are fused and may have spines or tubercules projecting from them.

There are ten especially important groups within the phylum Echinodermata: Homalozoa, Cystoidea, Blastoidea, Crinoidea, Asteroidea, Ophiuroidea, Helicoplacoidea, Edrioasteroidea, Holothuroidea, and Echinoidea.

Homalozoa is an extinct group of echinoderms which existed only from the Cambrian to the Devonian period. They do not show the typical radial symmetry of other echinoderms but instead are asymmetrical. A few groups show an imperfect bilateral symmetry. The body is composed of a <u>theca</u> which is somewhat flattened and covered with plates. A jointed brachial appendage called an <u>aulacophore</u> protrudes from the anterior of the body. The animal's mouth was probably located at the base of the aulacophore. The anus is at the opposite end of the body. Unfortunately very little is known of the habits or life-style of homalozoans since there are no living representatives.

Chapter 12: Echinodermata

Cystoids, blastoids, and crinoids all look very much alike with varying degrees of complexity. Cystoids, the most primitive, lived from the Ordovician to the Devonian period. They show imperfect radial symmetry and are shaped like a knob on the end of a stem. The stem is arranged like a stack of "life savers", each unit being called a columnal. The body or theca of the cystoid is made up of many plates perforated by small openings making up a pore rhomb. Appendages, called brachioles project from the top and were probably food gathering organs. Cystoids lived on the ocean floors and were presumably attached to the bottom by their stems.

Blastoids lived from the Ordovician to the Permian period and are more complex than cystoids. Radial symmetry is well defined, as indicated by five prominent channels or ambulacral grooves thought to be used in food transport. The calcareous plates making up the theca or calyx are more regularly shaped than those of cystoids. Openings or slits penetrate the plates, probably for respiration. The numerous brachioles used in food gathering are somewhat bushy in appearance. The stem is slender and much shortened, anchoring the blastoid to the substrate. In some species, the stem is completely absent, perhaps secondarily lost.

Crinoids, living from Ordovician to Recent times, are more complex than either cystoids or blastoids. The arms are finely branched feathery appendages. In fossil crinoids, the stems are larger than those of blastoids and cystoids and the columnals are arranged in a very regular pattern, some of them having lateral projections called cirri. Crinoids were an important part of Paleozoic reefs, reaching their zenith during the Mississippian period. Carbonate sediments from this Paleozoic period are rich in crinoid stems, reflecting their importance in those paleocommunities. Modern crinoids usually are without stems and live in oceans world-wide as free swimming invertebrates.

Class Asteroidea consists of starfish or sea stars. They occur in the fossil record from the Ordovician period to the present. Asteroids have well developed water vascular systems and radiating arms symmetrically arranged on the body. These usually are five in number, although more may be present, and show a unique ability to regenerate when broken.

Class Ophiuroidea, or brittle stars, somewhat resemble starfish but are much smaller. The arms are very mobile, slender structures, which often are numerously branched into a coiling mass. They extend from a flattened central disc which is quite distinct from the arms. The habits of brittle stars differ from the predatory starfish since they feed mainly on detritus and occasionally smaller animals.

Chapter 12: Echinodermata

Holothuroidea are the sea cucumbers. They do not have arms or stems. The body is elongated, showing a trend back to the bilateral symmetry of more primitive echinoderms. The skeletal plates, so prominent in other echinoderms, are reduced in holothuroids to calcium carbonate sclerites or ossicles embedded in the skin. These disarticulted sclerites are the only part of the holothuroids found in the fossil record and are not abundant. Consequently, the holothuroids are not well known as fossil species.

Helicoplacoidea are very distinctive forms of early echinoderms which lived only during the Cambrian period. The body lacks arms or stem and is spindle shaped with pleated spiral rows of plates and generally a single ambulacrum. Theoretically, the "pleating" allowed for expansion and contraction, a perhaps useful function in locomotion.

Edrioasteroidea, an extinct class, lack arms and are small button-like echinoderms having frequently sinuous ambulacra arranged to form a star shape on one surface. Tubercles on the plates indicate that they may have had spines of some sort, while special pores between the ambulacral plates suggest openings for podia. Although some edrioasteroids had stems, others did not. They lived a sedentary life from Cambrian to Mississippian times on ocean floors.

Chapter 12: Echinodermata

Class Echinoidea includes sand dollars and sea urchins. The calcium carbonate plates typical of the echinoderms are fused into a rigid skeleton from which moveable spines project. There are no arms or stem, only a spherical or flattened disc-shaped body showing radial symmetry. The five ambulacral grooves have lost their feeding function and are now located on the aboral rather than the oral surface. On the underside, within the mouth, is a calcareous chewing apparatus called Aristotle's Lantern. Echinoids have lived from Ordovician time to the present. Although infrequently found in the Paleozoic era, they are an important part of the fossil marine faunas in Mesozoic and especially Cenozoic time.

B. Synopsis

Phylum Echinodermata [Cambrian-Recent]

- Radial symmetry in adults
- Water vascular system
- Skeleton of calcium carbonate plates

Subphylum Homalozoa [Cambrian-Devonian]

- Flattened body or theca with jointed brachial appendage called an aulacophore

Subphylum Crinozoa [Cambrian-Recent]

Class Cystoidea [Ordovician-Devonian]

- Calyx and stem; imperfect radial symmetry; brachioles

Chapter 12: Echinodermata

Class Blastoidea [Silurian-Permian]

Pronounced radial symmetry; prominent ambulacra; bushy brachioles

Class Crinoidea [Ordovician-Recent]

Feathery brachioles; large stems with some columnals having cirri

Subphylum Asterozoa [Ordovician-Recent]

Class Asteroidea [Orovician-Recent]

No stems; arms with tube feet; radial symmetry

Class Ophiuroidea [Ordovician-Recent]

Slender, active brittle arms radiating from central disc

Subphylum Echinozoa [Cambrian-Recent]

Class Helicoplacoidea [Cambrian]

Spindle shape with pleated spiral rows of plates and ambulacrum

Class Edrioasteroidea [Cambrian-Mississippian]

Small disc with star-shaped ambulacra on top

Class Holothuroidea [Devonian-Recent]

Skeleton reduced to calcium carbonate sclerites or ossicles embedded in the skin

Class Echinoidea [Ordovician-Recent]

Skeletal plates fused, rigid, with spines

Aristotle's Lantern

Chapter 12: Echinodermata

C. Representative Genera

Placocystites (Homalozoa)

Silurian

Mitrocystites (Homalozoa)

Ordovician

Helicoplacus (Helicoplacoidea)

Cambrian

Edrioaster (Edrioasteroidea)

Ordovician

Isorophus (Edrioasteroidea)

Ordovician

Pleurocystites (Cystoidea)

Ordovician

Caryocrinites (Cystoidea)

Ordovician-Silurian

Pentremites (Blastoidea)

Mississippian-Pennsylvanian

Heteroschisma (Blastoidea)

Devonian

Eucalyptocrinites (Crinoidea)

Silurian-Devonian

Platycrinites (Crinoidea)

Devonian-Permian

Devonaster (Asteroidea)

Devonian

Chapter 12: Echinodermata

Hudsonaster (Asteroidea)

Ordovician-Silurian

Ophiura (Ophiuroidea)

Cretaceous-Recent

Ophioderma (Ophiuroidea)

Recent

Cucumaria (Holothuroidea)

Ordovician-Recent

Holaster (Echinoidea)

Cretaceous-Recent

Archaeocidaris (Echinoidea)

Permian-Carboniferous

ECHINODERMATA MORPHOLOGY

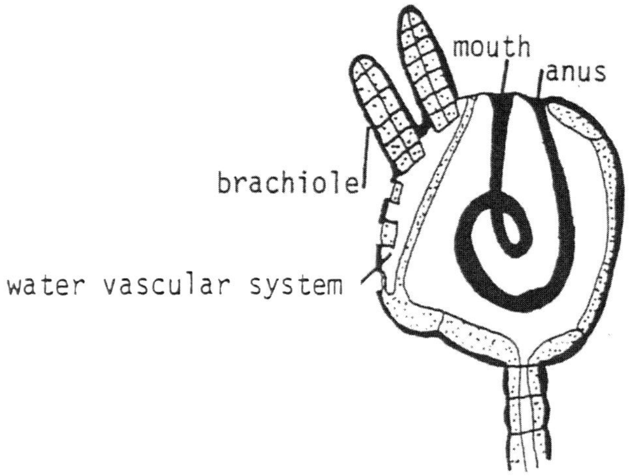

BLASTOID

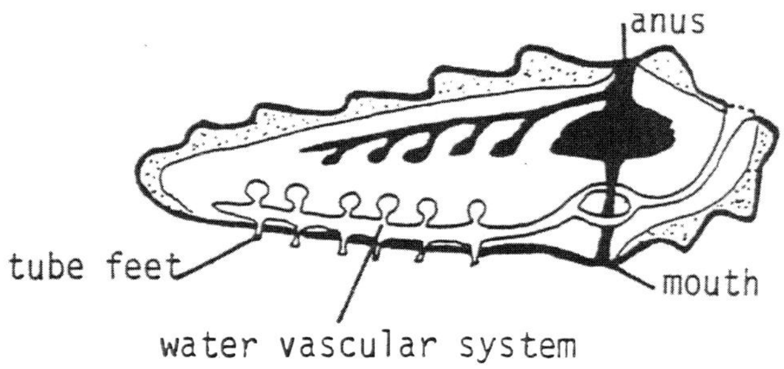

STARFISH

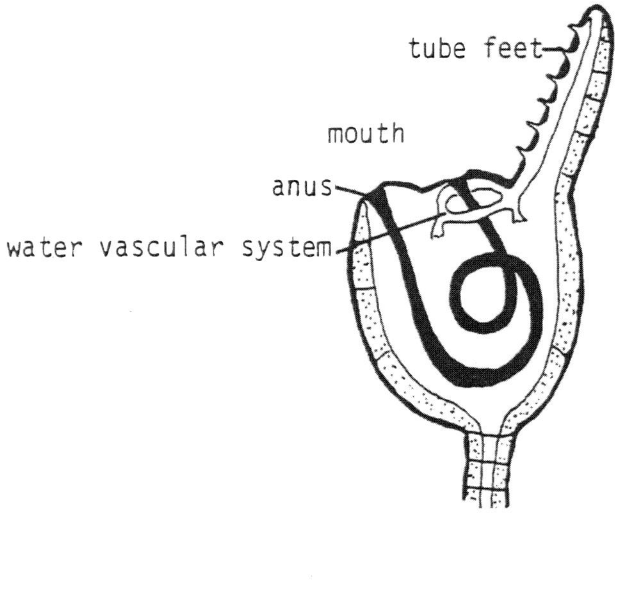

CRINOID

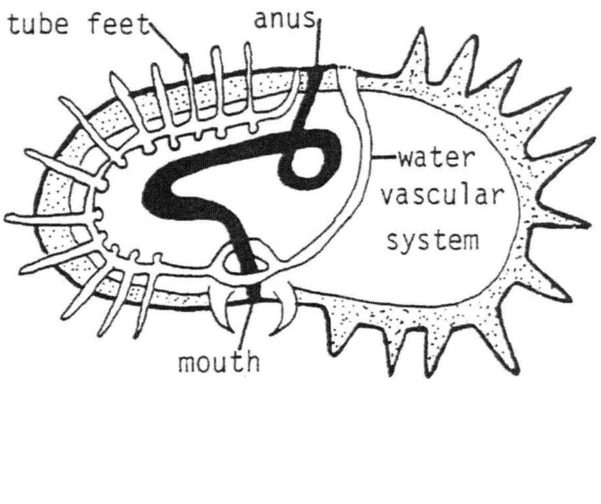

ECHINOID

ECHINODERMATA

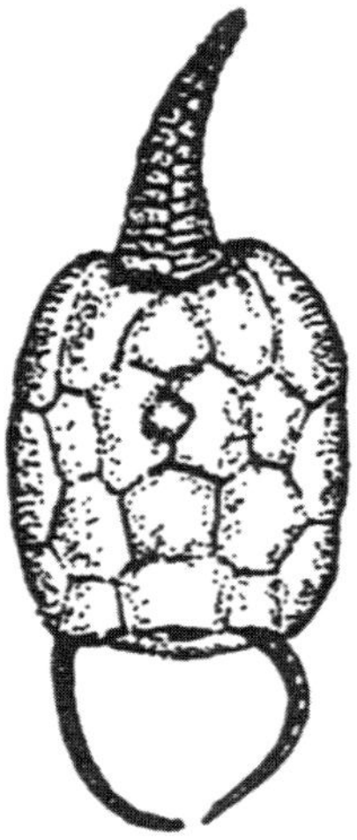

Placocystites

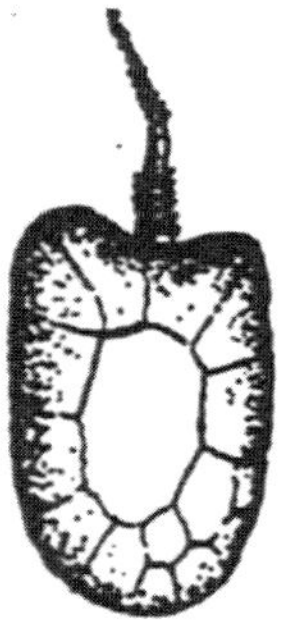

Mitrocystites

Helicoplacus

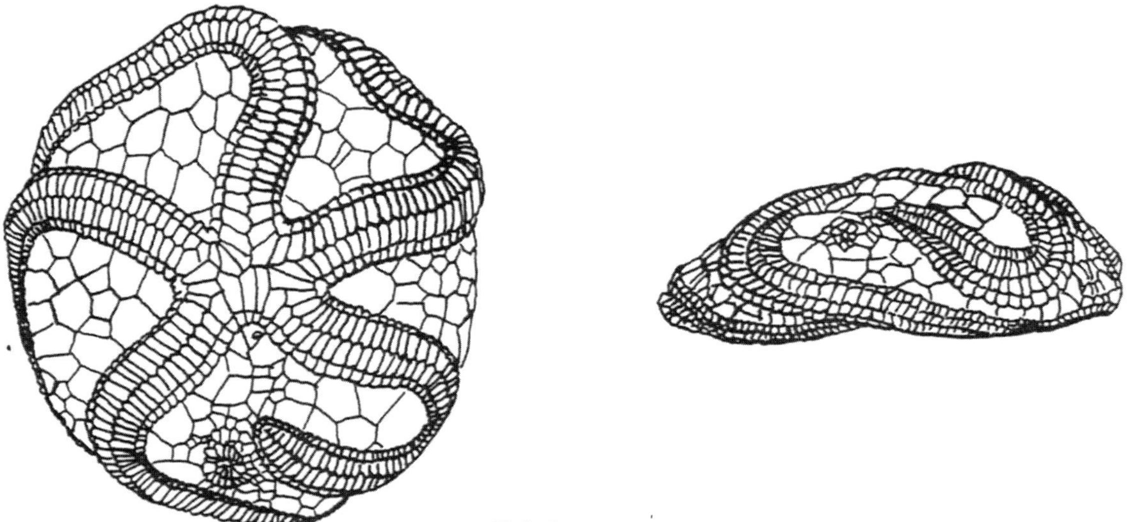

Edrioaster

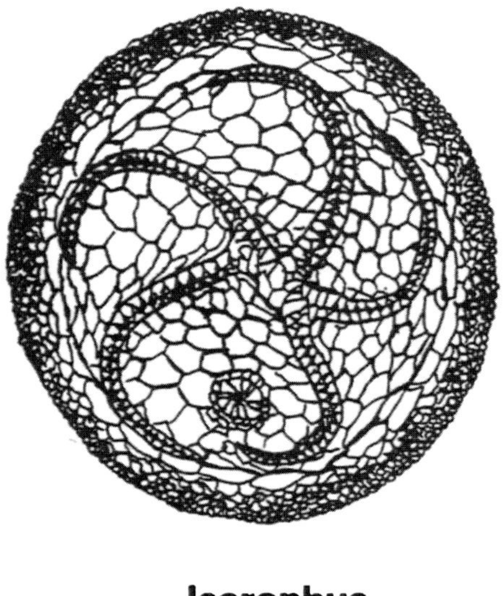

Isorophus

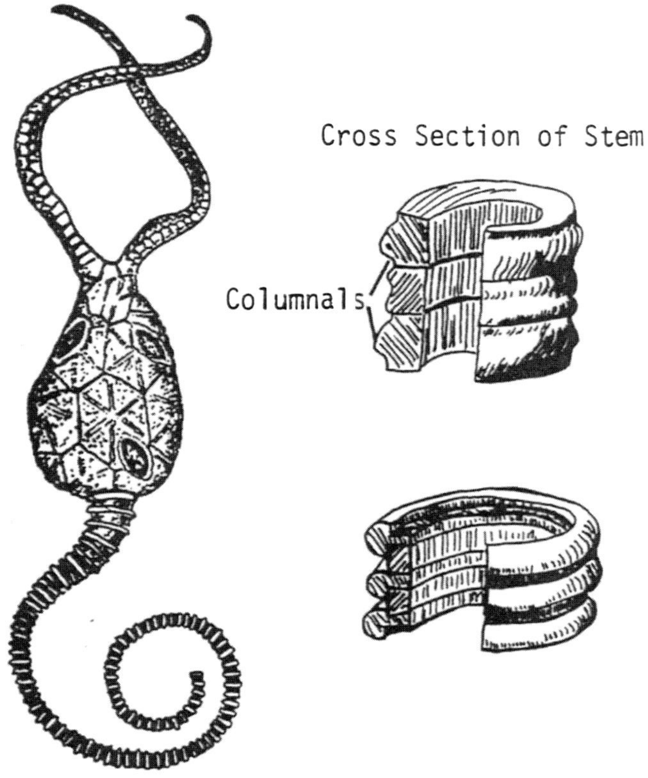

Pleurocystites

(After Treatise)

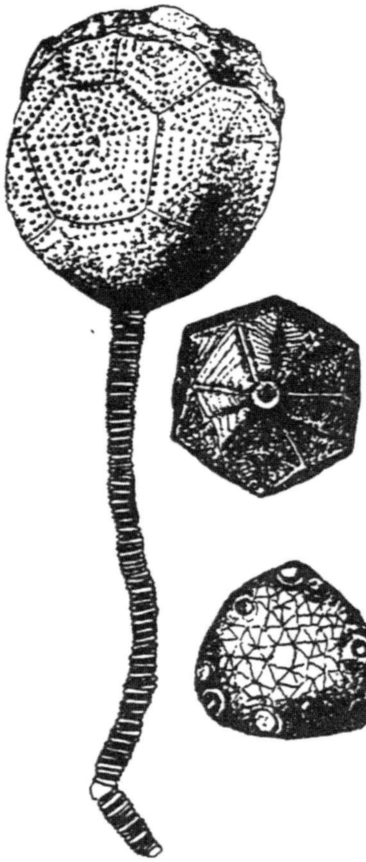

Caryocrinites

(After Treatise)

Pentremites

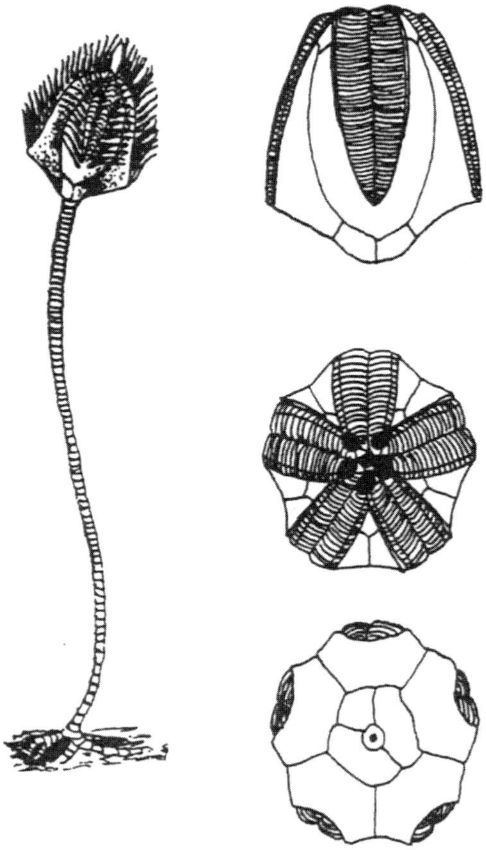

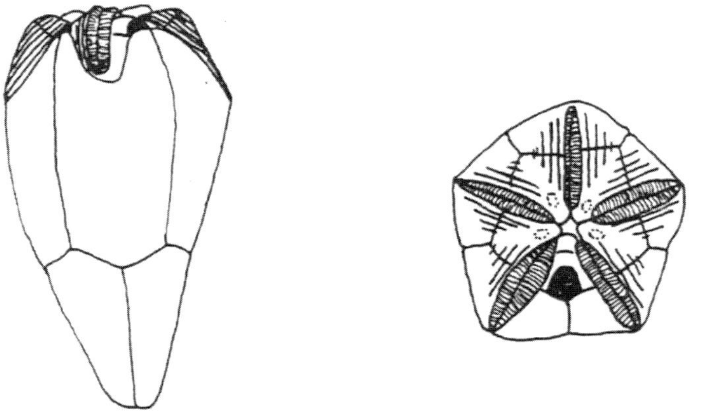

Heteroschisma

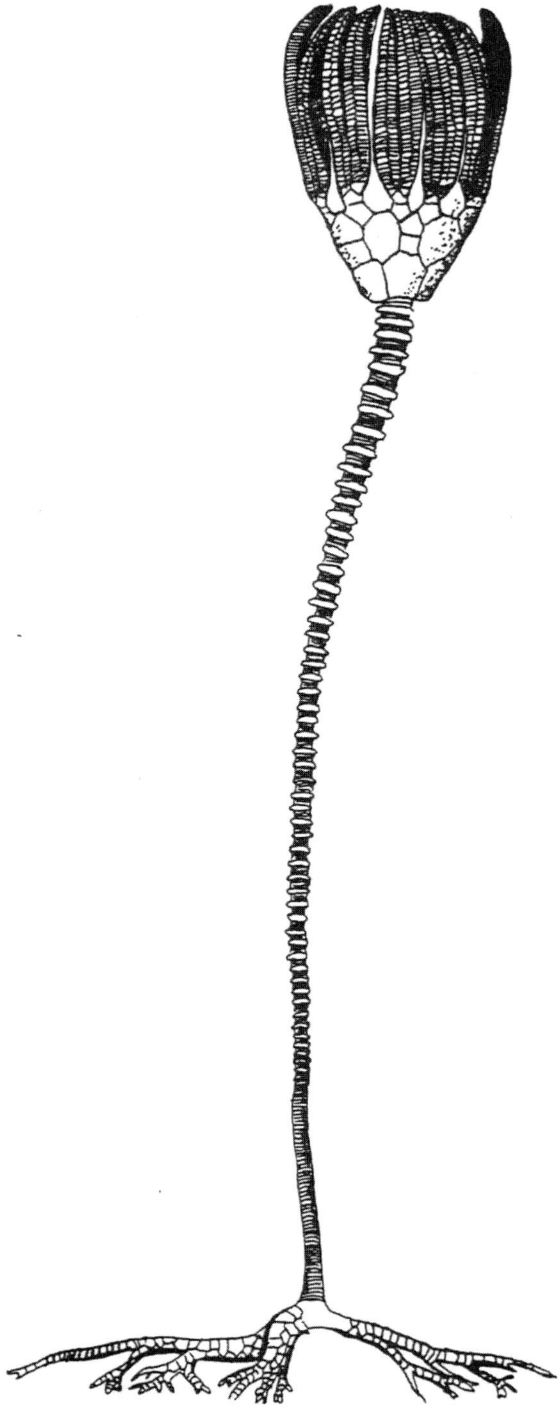

Eucalyptocrinites

crinoid column

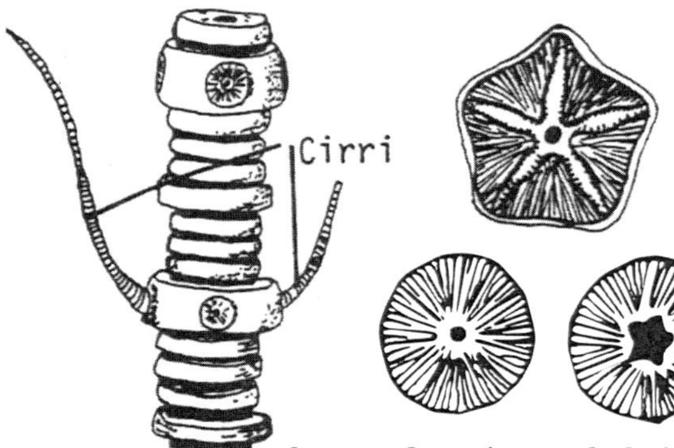

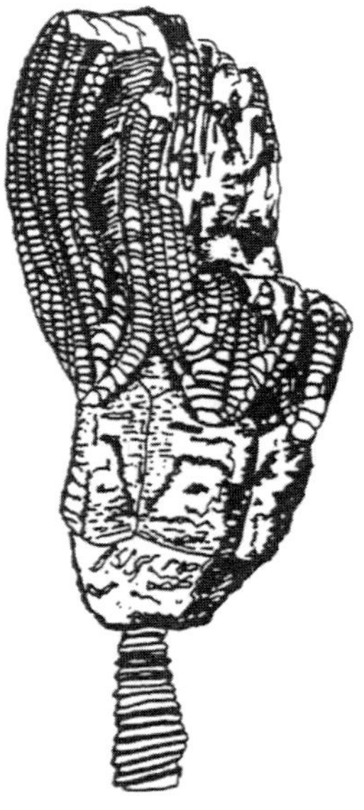

Platycrinites

(After Treatise)

Devonaster

Hudsonaster

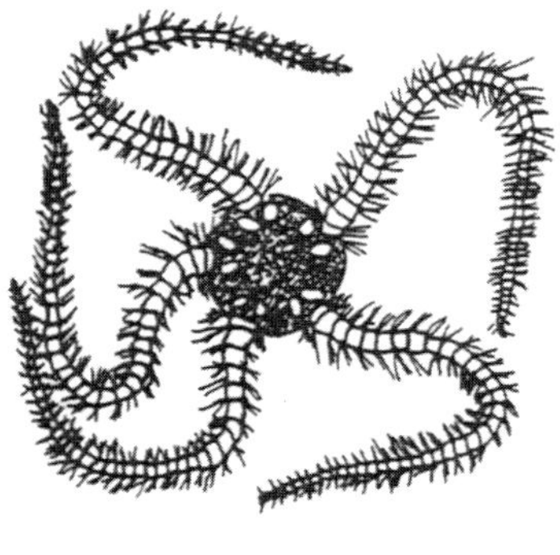

Ophiura

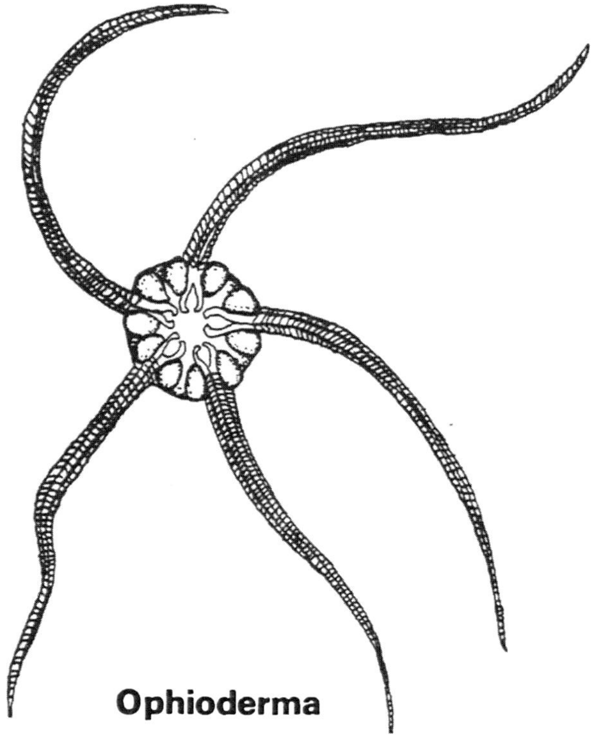

Ophioderma

Cucumaria

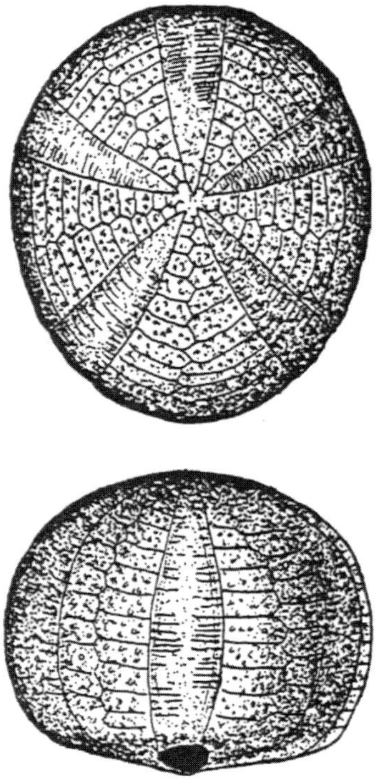

Holaster

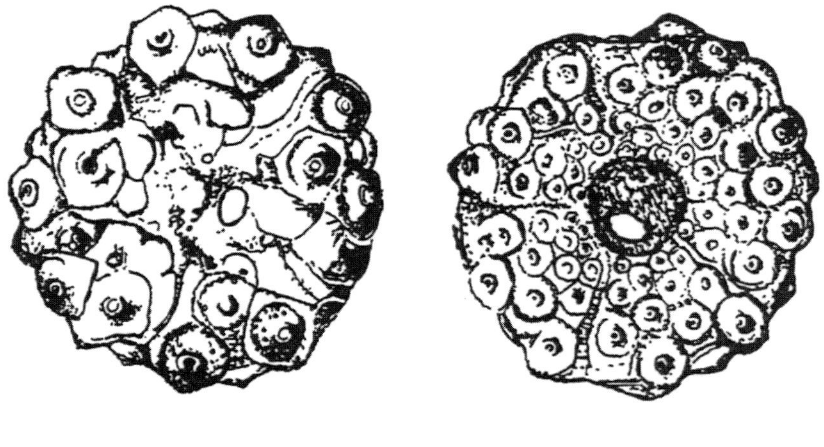

Archaeocidaris

NOTES

Chapter 13: GRAPTOLITHINA

A. Discussion

Class Graptolithina is a group of colonial marine organisms ubiquitously distributed from the Cambrian to the Carboniferous period. They are typically preserved as carbon films of leaf, twig, and web-like appearances found on black shale bedding planes. The detailed three dimensional structure has been interpreted from a few well preserved unflattened specimens found in limestone and flint.

The structural material of a graptolite colony was an insoluble fibrous protein called scleroprotein. In its simplest form, the colony formed a single or double branch, each of which is called a stipe, bearing small cup-like thecae. The organisms presumably lived inside the thecae and thrust their feeding apparati into the water somewhat like bryozoans. The stipes have characteristic growth bands and were attached to the ocean floor or a floating object by a thin thread called a nema. The entire colony, called a rhabdosome, is composed of stipes, thecae, and nema.

As graptolites evolved, the stipes became less numerous and rotated from a pendant or hanging position below the nema to an upward or "V" shaped position with the nema inside the stipes. By the Ordovician period the stipes of many species had fused in the upward position, enclosing the nema. Stipes which grow erect, either along or enclosing the nema, are termed scandent. The shape of the thecae also altered during the evolutionary process. Structural changes in the shapes of graptolites over time are so pronounced and diagnostic for certain species that graptolites are important index fossils for Paleozoic rocks.

There are numerous orders of Graptolithina; however, two orders contain the majority of specimens. These are the orders Dendroidea and Graptoloidea. Dendroidea were sessile and usually attached to a substrate. They had a characteristic thecal grown patten which gave the colony a bushy appearance. Although the Dendroidea extended from the Cambrian to the Carboniferous period, they are considered the most primitive of the graptolites.

Graptoloidea were floating colonies, possibly attached to planktonic seaweeds. They are considered to be more advanced than the Dendroidea because they show reduction in the number of stipes and thecae. In some species the stipes have become fused into biserial forms or double rows of thecae. The Graptoloidea are excellent index fossils for the Ordovician to Devonian periods because they were abundant, geographically widespread, and had short well defined biostratigraphic ranges.

B. Synopsis

Class Graptolithina [Cambrian-Carboniferous]

Colonial marine organisms with stipes, thecae and nema

Order Dendroidea [Cambrian-Carboniferous]

Bushy sessile colonies

Order Graptoloidea [Ordovician-Devonian]

Floating colonies with reduction in the number of stipes and thecae

GRAPTOLOIDEA

graptoloid morphology

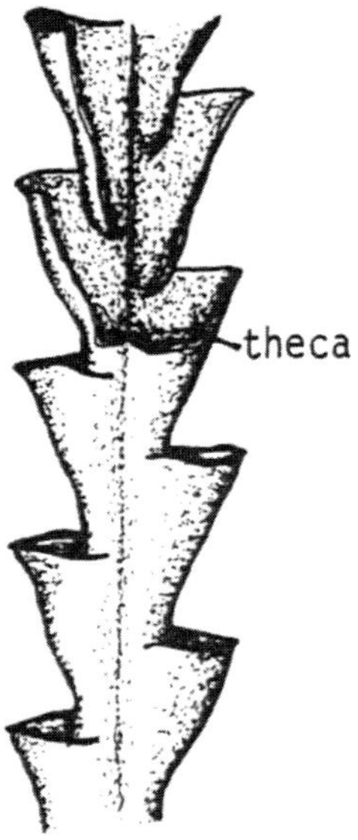

stipe

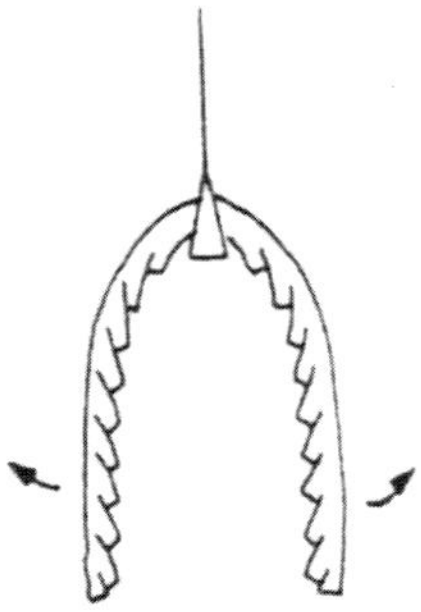

pendant

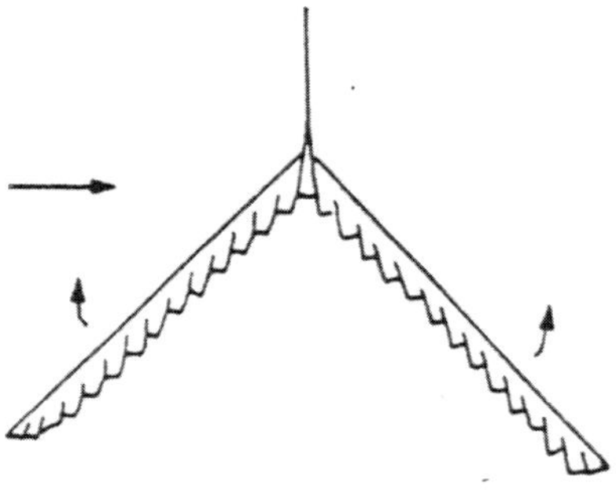

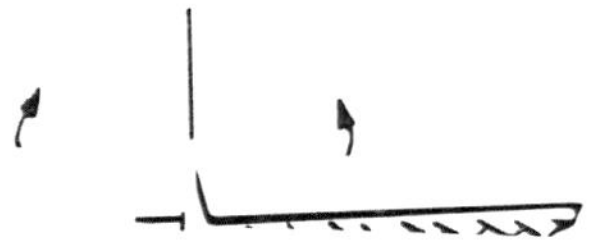

horizontal

evolution of graptoloids from
uniserial to biserial colonies

scandent

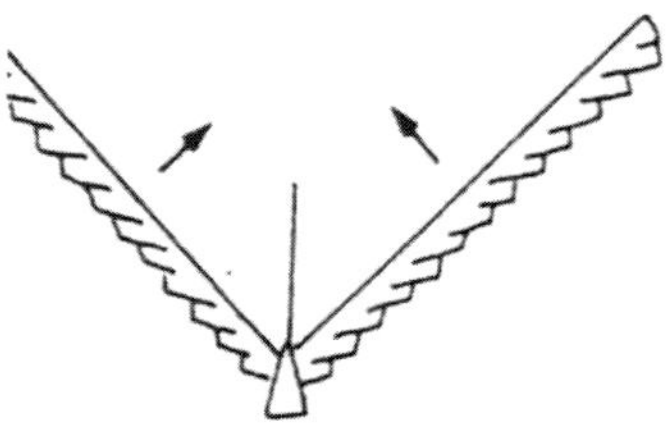

GRAPTOLOIDEA

Monograptus
(Silurian)

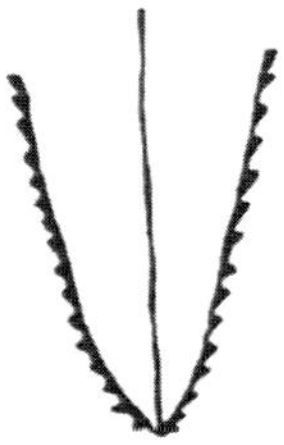

Dicellograptus
(Middle to Late Ordovician)

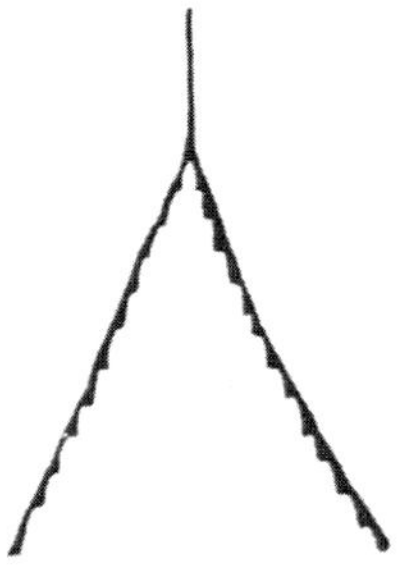

Didymograptus
(Early to Middle Ordovcian)

Diplograptus
(Middle Ordovician to Late Silurian)

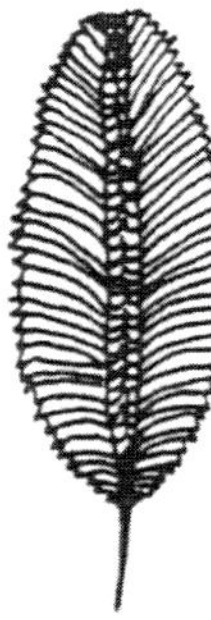

Phyllograptus
(Late Ordovician)

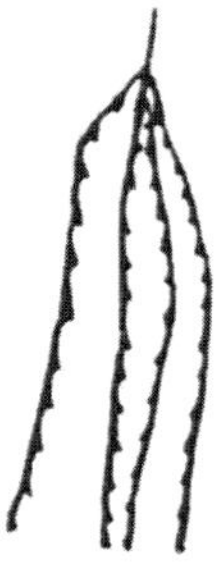

Tetragraptus
(Early Ordovician)

DENDROIDEA

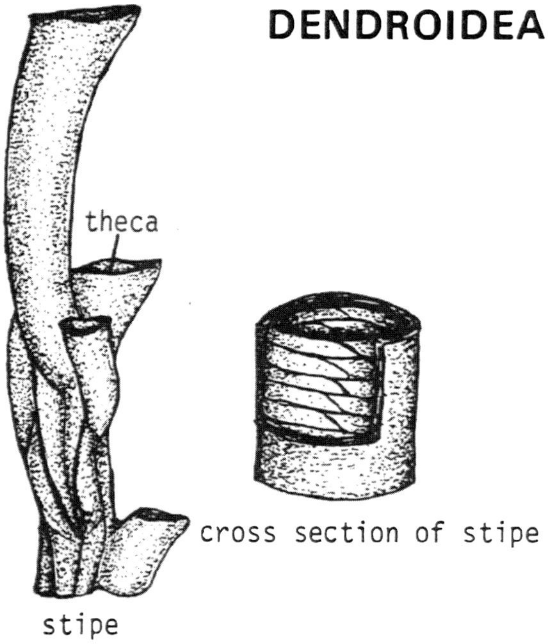

dendroid morphology

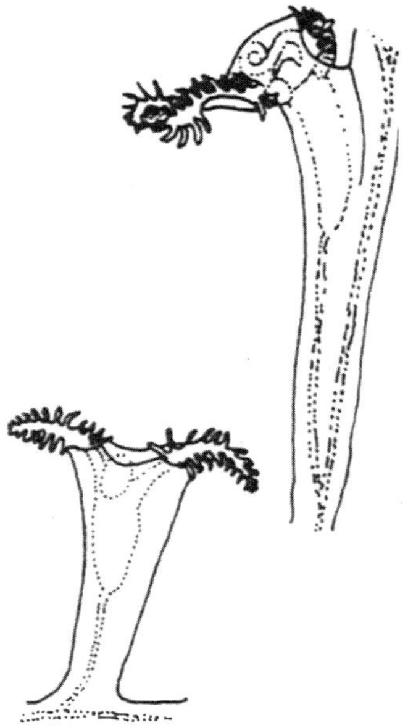

possible relationships of animal to exoskeleton

Ptiograptus

Licnograptus

Dendrograptus

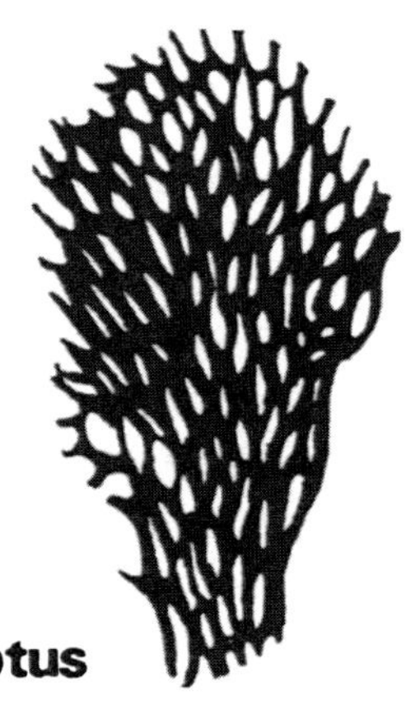

Desmograptus

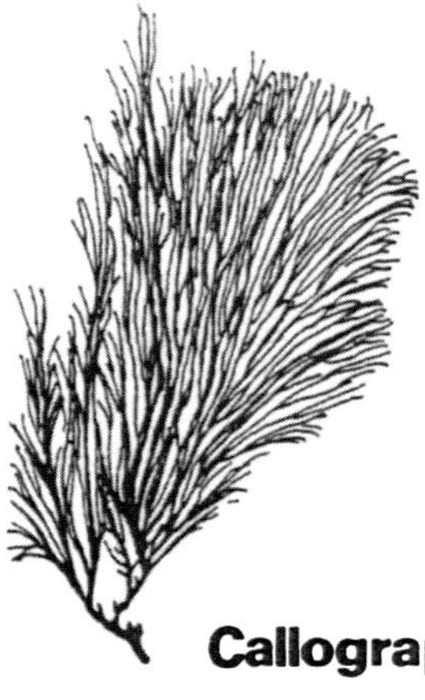

Callograptus

NOTES

Chapter 14: CONODONTA

A. Discussion

Phylum Conodonta contains Paleozoic microfossils, resembling teeth, which belong to an abundant but very rarely preserved organism. They are composed of calcium phosphate which was readily preserved. As a consequence, they occur abundantly throughout the world and have been very useful for both correlation and dating of sedimentary rock from Cambrian to Triassic times.

Conodonts are classified by their shape into three categories: coniform, ramiform, and pectiniform. Coniform elements are single tusk or tooth shaped microfossils which are found in Cambrian to Silurian rocks. Ramiform elements have a laterally extended base, which may be forked, and has denticles or small teeth jutting from the surface. Ramiform conodont elements are important index fossils for Ordovician to Triassic strata. Pectiniform elements have a plate or platform from which denticles protrude. They are very useful for dating Ordovician to Triassic sedimentary rocks.

As microfossils, conodonts are found singly or in assemblages. Assemblages frequently have right and left handed members which indicate that the original organism was bilaterally symmetrical. Recent discovery of a real conodont animal with an assemblage indicates that the conodonts in the living organism probably functioned in gathering food and helping the animal to swallow it. Older theories suggested that the conodonts were the support structures for some as yet unidentified organ or perhaps a filtering device.

B. Synopsis

Phylum Conodonta [Cambrian-Triassic]

Teeth-like microfossils

Classified as coniform, ramiform, and pectiniform

CONODONTA

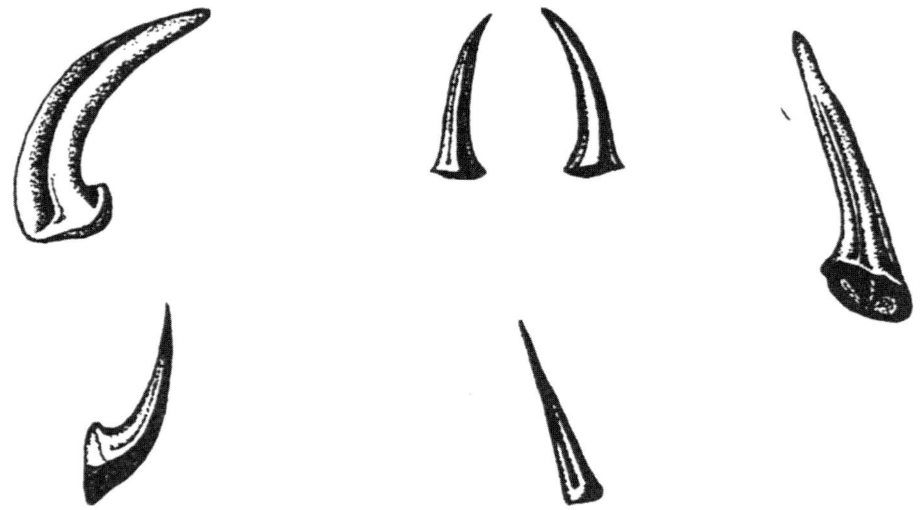

coniform

Longitudinal Section Demonstrating Lamellar Mode of Growth in Coniform Conodont

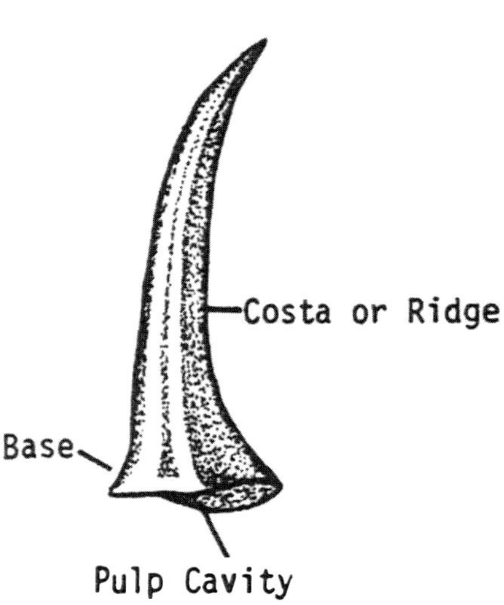

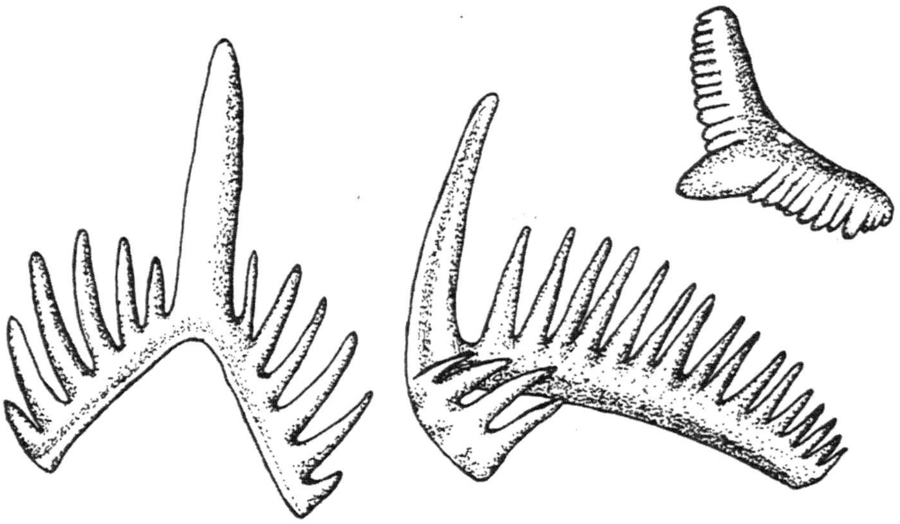

ramiform

pectiniform

NOTES

ADDITIONAL REFERENCES

1. Barnes, Robert D. 1968. Invertebrate Zoology. Philadelphia: W. B. Saunders.

2. Beerbower, James R. 1968. Search for the Past. New Jersey: Prentice-Hall.

3. Clarkson, E.N.K. 1979. Invertebrate Paleontology and Evolution. London: Allen & Unwin.

4. Encyclopedia of Paleontology. The Encyclopedia of Earth Sciences, Volume VII. 1979. Stroudsburg: Dowden, Hutchinson & Ross.

5. Hyman, Libbie H. 1940-67. The Invertebrates. 6 Vols. New York: McGraw-Hill.

6. Moore, Raymond C.; Lalicker, Cecil G. and Fisher, Alfred G. 1952. Invertebrate Fossils. New York: McGraw-Hill.

7. Moore, Raymond C. (founding ed.). 1953-present. Treatise on Invertebrate Paleontology. Geo. Soc. Am. and Univ. of Kansas, Lawrence.

8. Nitecki, Matthew H. 1972. North American Silurian Receptaculitid. Fieldiana, Geology, Volume 28.

9. Shimer, Hervey W. and Shrock, Robert T. 1944. Index Fossils of North America. New York: John Wiley and Sons.

10. Synopsis and Classification of Living Organisms. 1982. McGraw Hill: New York.

11. Weller, J. Marvin. 1969. The Course of Evolution. New York: McGraw-Hill.